The Differences Between Intensive and Extensive Fish Farming

THE DIFFERENCES BETWEEN INTENSIVE AND EXTENSIVE FISH FARMING

Edited by:

Navghan Mahida

Delve
PUBLISHING

www.delvepublishing.com

The Differences Between Intensive and Extensive Fish Farming

Navghan Mahida

Delve Publishing
224 Shoreacres Road
Burlington, ON L7L 2H2
Canada
www.delvepublishing.com
Email: orders@arclereducation.com

ISBN: 978-1-77469-428-2 (Hardcover)

ABOUT THE EDITOR

Dr. Navghan Mahida is a self-motivated fisheries scholar by qualification with over five years of work experience, including over 9 training programs and work exposure, possesses qualities of leadership, good communication & interpersonal skills, to do attitude, and ability to quickly adapt new environments. He is working as Assistant Professor in the Department of Animal Husbandry, Dairy and Fisheries, Parul University Vadodara-Gujarat. He had completed B.F.Sc (Bachelor of Fisheries Science) in 2012 from CoF-Veraval, Junagadh Agricultural University, Junagadh. He pursued M.F.Sc in 2014 and Ph.D. in 2019 from the Central Institute of Fisheries Education-Mumbai. He started his career as Fisheries Officer in the Government of Gujarat in 2013 and further worked in MNC Shoreline Aquaculture as Assistant Manager. He specialized in fisheries economics with commendable academic records and achievements. He has published 25 research papers in national and international peer-reviewed journals so far, including conduction and participation in conferences, conventions, symposiums, seminars, training, etc. He had written souvenirs for the students' convention programs and three popular articles. He is a lifetime member of fisheries and life sciences, the European Journal of fisheries. He is also a reputed editor in the journal of fisheries and life sciences, PLOS ONE San Francisco, California, USA, Asian Journal of Economics, Extension and Social Science Thailand etc.

TABLE OF CONTENTS

List of Figures ..... *xi*

List of Abbreviations ..... *xv*

Introduction ..... *xvii*

Preface ..... *xix*

Chapter 1 Introduction to Intensive and Extensive Fish Farming 1

1.1. Introduction 2

1.2. Extensive Fish Farming 3

1.3. Aquaculture 7

1.4. Reservoir Development 9

1.5. Water Quality 11

1.6. Intensive Fish Farming 12

1.7. Fish Production 14

1.8. Challenges 16

1.9. Hydroponics 17

1.10. Nitrification 18

1.11. Growth 20

1.12. Factors 22

1.13. Legislation 25

Chapter 2 Intensive Fish Farming 27

2.1. Introduction 28

2.2. Culture Systems 30

2.3. Foods 31

2.4. Intensive Systems 32

2.5. Problems 34

2.6. Aquaculture Sector 35

2.7. Nutritional Requirements 37

2.8. Aquatic Environment 38
2.9. Fish Appetite 41
2.10. Receptors 43
2.11. Freshwater Fish 46
2.12. Analysis 50

Chapter 3 Financial Management and Economics of Fish Farming 55
3.1. Introduction 56
3.2. Business 57
3.3. Marketing Problems 59
3.4. Evaluation 60
3.5. Consumer Demand 62
3.6. Restrictions 65
3.7. Credit Reserves 71

Chapter 4 Extensive Fish Farming 75
4.1. Introduction 76
4.2. Strategies 76
4.3. Treatment 78
4.4. Combined Intensive-Extensive (CIE) System 79
4.5. Nitrification 81
4.6. Polyculture 83
4.7. Crabs 84
4.8. Shrimp 86

Chapter 5 Best Management Practices in Fish Farming 89
5.1. Introduction 90
5.2. Aquaculture 91
5.3. Best Management Practices (BMPS) 94
5.4. Current Trends 95
5.5. Application 97
5.6. Waste Management 98
5.7. Restrictions 100
5.8. Environmental Problems 102
5.9. Initiatives 103

5.10. Development 105
5.11. Watershed Ponds 107
5.12. Brackish Water 112
5.13. Commercial Production 119

Chapter 6 Sustainable Fish Farming 121
6.1. Introduction 122
6.2. Oyster 122
6.3. Water Quality 123
6.4. Mechanism 125
6.5. Food Production 126
6.6. Microbes 132
6.7. Sustainability 134
6.8. Nutrition 135
6.9. Microalgae 136
6.10. Natural Resources 138
6.11. Diseases 143
6.12. Ammonia 144

Chapter 7 Impact of Climate Change on Fish Farming 151
7.1. Introduction 152
7.2. Marine Fisheries 152
7.3. Climate Change 155
7.4. Capture Fisheries 158
7.5. Sea Surface Height 159
7.6. Breeding Grounds 162
7.7. Temperature And Salinity 167

Chapter 8 Recirculating Fish Farming 175
8.1. Introduction 176
8.2. Ponds 177
8.3. Recirculating Systems 178
8.4. Hazards 180
8.5. Hardness 181
8.6. Water Flow 183
8.7. Diseases 186

Chapter 9 Reproductive Aquaculture 191

9.1. Introduction 192

9.2. Ovary 193

9.3. Reproduction 195

9.4. Spawning 198

9.5. Sodium Ions 201

9.6. Sperm Motility 202

9.7. EGG Quality Management 208

Index 213

LIST OF FIGURES

Figure 1.1. Extensive and intensive fish farming
Figure 1.2. Homestead pond
Figure 1.3. *Labeo calbasu*
Figure 1.4. Aquaculture in India
Figure 1.5. Reservoir development
Figure 1.6. Water quality
Figure 1.7. Intensive fish farming
Figure 1.8. Polyculture
Figure 1.9. Rice and fish farming
Figure 1.10. Hydroponics system
Figure 1.11. Nitrification
Figure 2.1. Intensive fish farming
Figure 2.2. Giant tiger shrimp
Figure 2.3. Phytoplankton
Figure 2.4. Shrimp aquaculture
Figure 2.5. Hypothalamus
Figure 3.1. Aquaculture enterprises
Figure 3.2. Fingerlings
Figure 4.1. Intensive tilapia farming
Figure 4.2. *Penaeus vannamei*
Figure 4.3. Nitrifying bacteria
Figure 4.4. Biofilter
Figure 4.5. Australian crayfish
Figure 4.6. Mud crabs
Figure 4.7. *Brachionus*
Figure 5.1. Capture fisheries
Figure 5.2. Grass carp
Figure 5.3. Silver carp

Figure 5.4. Bighead carp
Figure 5.5. Waste
Figure 5.6. Embankment ponds
Figure 5.7. Watershed ponds
Figure 5.8. Crayfish culture
Figure 5.9. Evapotranspiration
Figure 5.10. El Nino Southern oscillation
Figure 5.11. Net pens
Figure 5.12. Trawlers
Figure 5.13. Yellow perch
Figure 6.1. Natural oyster reefs
Figure 6.2. *Mugil cephalus*
Figure 6.3. *Chirostoma estor*
Figure 6.4. *Cichlasoma urophthalmus*
Figure 6.5. *Herinchthys cyanogutattus*
Figure 6.6. *Goodea atripinis*
Figure 6.7. *Xenotoca variata*
Figure 6.8. *Goodea gracilis*
Figure 6.9. *Girardinichthys multiradiatus*
Figure 6.10. *Macrobrachium rosenbergii*
Figure 6.11. Biofloc systems
Figure 6.12. *Apocyclops dengizicus*
Figure 6.13. Microalgae
Figure 6.14. *Debaryomyces hansenii*
Figure 6.15. Pyrosequencing
Figure 6.16. Hyperplasia of gill tissue
Figure 7.1. Marine capture fisheries
Figure 7.2. GCM
Figure 7.3. Sea surface height
Figure 7.4. Five-ray yellowtail
Figure 7.5. *Seriola quinqueradiata*
Figure 7.6. Spiny lobster
Figure 7.7. Spanish mackerel
Figure 8.1. Recirculating aquaculture systems

Figure 9.1. Reproductive aquaculture
Figure 9.2. Artificial insemination
Figure 9.3. Cryopreservation
Figure 9.4. *Macrozoarces americanus*

LIST OF ABBREVIATIONS

AcP	acid phosphatase
AspAT	aspartate aminotransferase
BMPs	best management practices
BOD	biochemical oxygen demand
CIE	combined intensive-extensive
DGGE	denaturing gradient gel electrophoresis
DHA	docosahexaenoic acid
DO	dissolved oxygen
EAC	east Australian current
EDCs	endocrine-disrupting chemicals
ENSO	El Nino Southern oscillation
EPA	eicosapentaenoic acid
EU	European Union
FAO	Food and Agricultural Organization
HUFA	highly unsaturated fatty acids
IPCC	intergovernmental panel on climate change
LD	light/dark
LDH	lactate dehydrogenase
LT	low temperature
N	nitrogen
ODAS	organic detrital algae soup
OM	organic matter
P	phosphorus
PDO	Pacific decadal oscillation
PUFAs	polyunsaturated fatty acids
RAS	recirculating aquaculture system
SCP	single cell proteins
SDA	specific dynamic action
SOD	superoxide dismutase

SST	sea surface temperature
TAN	total ammonium nitrogen
TSS	total suspended solid
TTGE	temperature gradient gel electrophoresis
USEPA	United States Environmental Protection Agency
UV	ultraviolet

INTRODUCTION

The fish farming industry has been showing progress in the last decades, and this can be due to increased population growth and nutritional requirements. The extensive and intensive farming system includes pond systems, irrigation ditches, natural furrows, cage systems, fry farming establishments, and integrated recycling systems. Intensive fish farming is being practiced by many farmers due to its immediate benefits at the expense of natural resources. Shrimp culture in the terrestrial lands is an example of intensive fish farming that has changed the financial status of the farmers. The salmon and trout can be grown on a large scale to limit the losses. In order to meet the population demands, it is important to opt for best management practices (BMPs) in the aquaculture sector. Urbanization has produced immediate demand for healthy food that can be obtained from the fishes and related products. This has loaded the environment with diverse waste materials. This needs to be avoided in order to maintain sustainable fish farming.

The reproductive aquaculture constitutes several areas of fish culturing, and it is a crucial area that needs to be focused on before the installation of the fish farm and during operation. Intensive farming demands higher capital costs, and the profits gained can be high considering some of the risks associated with it. Extensive fish farming is vulnerable to diverse environmental forces that can effectively lower the yields. Climate change can be classified as a key factor in defining the economic benefits of fish farms. This can be applied to both inland and offshore farms established under private and public partnerships. The hazard and disease management must be monitored on a regular basis to forecast the possible infections to the fish populations under culture. Water quality is yet another crucial determining factor that will change the financial dynamics easily.

The role of the trained professionals in conducting training programs in operating fish farms is important. The government-associated learning programs can easily approach the farmers, and they can reap benefits from it. The management plan for the installation of these farms is a crucial stage that requires expertise in the subject. The farmers need to rely on authentic professionals and follow the steps proposed. The combination of both intensive and extensive fish farming can be used for improved yields.

PREFACE

Fish farming is considered as an important solution for the population demands that are increasing in a geometric progression. Fish is considered as a healthy food, and its culture is essential across several regions. The primary reason for the development of fish farming firms across the world can be attributed to its low cost and limited maintenance. Intensive fish farming can be considered as a procedure wherein the aquaculture farm is supported by the external food supply and an efficient water supply and purification system. The intensive fish farming system can be broadly classified into the integrative recycling system, fry farming, and cage system. Extensive fish farming is powered by a self-sustaining system and an efficient regulating system. This farming can be seen in the ponds and irrigation ditches. These two fish farming systems can be combined to get better results, and it can be called combined fish culture. Some of the fish farming systems include floating constructions and cages wherein fish will be cultured using natural methods. Extensive fish farming is mostly based on natural resources, and with proper improvement in the techniques, the yields can be enhanced. The intensive fish farming system is expensive, and it needs a lot of effort to keep the business aloft. Several fish species, including tilapia and shrimp, are being cultured in these systems with various levels of benefits to the farmers and consumers. The nutritional requirements of the fish farming system are always important, and it needs to be handled by experienced personnel at all times. It defines the success of the fish farming business, and the disease management in these farms can be considered as a crucial element.

The financial management in intensive and extensive fish farming must be monitored by the farmers to forecast the yields. The mitigation measures should be planned in advance to minimize the losses incurred due to uneven growth rates or natural calamities. Extensive fish farming depends on several factors, and it is always vulnerable to environmental changes. Fish farming must be sustainable and should not use or leave harmful waste in the environment. The overexploitation of the inland regions for the sake of intensive farming should be avoided by the local authorities and farmers too. Climate change is also affecting the fish industry, and it has lowered the yields drastically.

This book provides an introduction to intensive and extensive fish farming, and it can be used by undergraduates, postgraduates, and research students who wish to work in the fish farming sector.

CHAPTER 1

INTRODUCTION TO INTENSIVE AND EXTENSIVE FISH FARMING

CONTENTS

1.1. Introduction 2
1.2. Extensive Fish Farming 3
1.3. Aquaculture 7
1.4. Reservoir Development 9
1.5. Water Quality 11
1.6. Intensive Fish Farming 12
1.7. Fish Production 14
1.8. Challenges 16
1.9. Hydroponics 17
1.10. Nitrification 18
1.11. Growth 20
1.12. Factors 22
1.13. Legislation 25

1.1. INTRODUCTION

Fish is an inexpensive and simply digestible animal protein, and it has been consumed for millennia from natural sources. Farmers can easily start raising fish in village ponds, tanks, or any other new water body, significantly improving their financial situation. The technology developed for fish culture allows multiple varieties of suitable fish to be cultivated at the same time. Composite Fish Culture is the most widely practiced method of fish culture in the country. Even seasonal ponds can be used for fish culture for a limited time. The area accessible for warm freshwater aquaculture under tanks and ponds is expected to be 2.85 million ha. Furthermore, 0.78 million hectares of swamps and other low-lying water-logged areas that are unsuitable for cultivation, as well as any agricultural land, can be transformed into ponds for fish farming (Tacon, 1995). Only about 20% of the possible area of tanks and ponds available for development has been placed under scientific fish culture, indicating enormous potential for horizontal expansion of composite fish culture (Prakash et al., 2018). The aquaculture sector accounts for around 60% of total inland fish production. This demonstrates the enormous potential for fish farming. Farmers demonstrated that by intensifying culture procedures and improving farm management, production levels might be improved by a factor of two, to roughly 5 tons per hectare per crop, and two crops can be grown in a year. Modern cages are floating constructions with a net suspended beneath them. Floating cages can be small and weak, or they can be massive, containing thousands of cubic meters of water and designed for usage in the open sea. Cages are used in fish cultivation during the grow-out phase, which lasts months or years until they are ready to sell.

In shallow water, such as ponds, pens, and net enclosures or hapas are utilized to establish a controlled environment for the culture of fish and crustaceans. They are usually small, measuring a few tens of square meters or fewer. Closely spaced stakes, such as bamboo stems or mangrove branches, or wire and other mesh can be used as enclosure walls. This culture is mostly practiced in poor countries. The use of mesh fences or walls to cage bottom-dwelling scallops is an interesting exception to the shallow water pen. The scallops will not be able to swim over the wall because of the height of these pens. To allow the mesh to raise and fall with the tide, they may utilize floats. Hapas are net enclosures constructed of fine nylon, plastic mosquito netting, or cotton mesh. Haps are incredibly easy to manage, and harvesting is considerably easier because the fry cannot escape. Hapas come in a variety of sizes, ranging from 1 to 40 m^3, with a depth of 1–2 m and are suspended on poles (Soltan, 2016). Simple squares or rectangles can be used. There are even more intricate designs, like those that use a succession of netting to separate broodstock and fry.

1.2. EXTENSIVE FISH FARMING

Extensive fish farming refers to fish farming in medium to large-sized ponds or water bodies where the fish production is solely dependent on natural productivity, which is only marginally or moderately improved. Inputs from outside sources are limited, costs are kept low, capital expenditure is minimal, and the number of fish produced per unit area is modest. In summary, production factor control is kept to a minimum. The labor return is high. In contrast, intensive fish farming indicates that a large number of fish are produced per unit of raising area. Production parameters including feed, water quality, and the quality of stocked fingerlings are all regulated to improve the culture's production conditions. Throughout the production cycle, there is constant monitoring. All of these controls, of course, imply high-tech techniques and capital-intensive investments, which raise manufacturing costs. Increasing production costs must be justified by increased profits. Natural productivity makes a minor contribution to fish output. Aside from that, intensive fish farming is associated with high expenses and environmental risks (Boyd et al., 2020) (Figure 1.1).

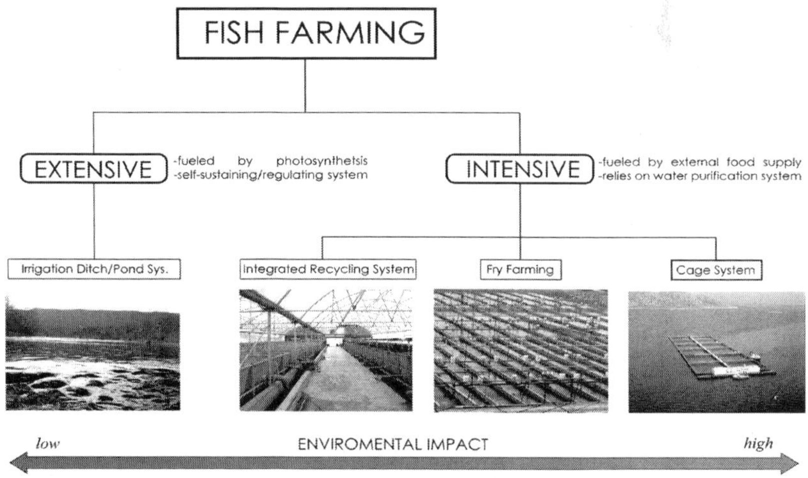

Figure 1.1. Extensive and intensive fish farming.

Source: http://1.bp.blogspot.com/-UTECtc7wvzc/UoB-eN2ouWI/AAAAAAAAACg/xhplMcTIAJk/s1600/25.jpg.

Apart from these two types of fish farming, some people talk about semi-intensive fish farming, which refers to techniques that include components of both. However, the term is vague. Before recommending one of these forms, it is necessary to consider the circumstances in which fish farming is carried out. The primary goal is to ensure food security and livelihood. This can be achieved by planting food crops first, and then diversifying the farmer's operations across a variety of agriculture and non-agriculture ventures in order to pursue his/her income-generating techniques. The rationale is to manage risk and deal with loss. Assets and investment capacities will have to be considered in the plans devised. Enabling market circumstances must exist for farmers to carry out strategies and achieve goals. Essentially, these are market systems for agricultural and non-agricultural factors and products. Food or labor shortages, for example, can affect market performance and prevent the farmer from attaining goals. In order to support objectives, the farmer seeks to maximize the utilization of resources within his grasp by utilizing unused or underutilized resources. The predicted return on investment in fish farming is thus the objective rationale for entering the industry. The decision to engage in income-generating activities is based on the projected returns and risks associated with each alternative.

Farmers can be split into three groups: those with few resources and land, limited or no cash, and a scarcity of labor, who are risk-averse and strive to diversify their production to distribute risk, but have limited maneuvering range; those with more resources, both on and off the farm; and those who are wealthy. The first group of farmers will use fish farming as a complement or supplementary activity, whereas the other two groups can see fish farming as a standalone source of revenue. This means that fish farming in the first group must be integrated into a larger farm system, whereas in the other two circumstances, it can stand alone and grow into a primary economic activity. The larger the market must be to absorb the created supply, the higher the function of fish production in generating money. The larger the domestic consumption of the fish produced; the fewer fish are produced. Except for the lowest, many smallholder farmers have off-farm incomes like selling, handicraft production, basket weaving, herbalist operations, wine, or beer manufacturing, and so on. These earnings are frequently significant since they are part of a risk-spreading strategy for securing a living. Farmers will increasingly rely on off-farm income to tide them over when agricultural production fails due to climate or other factors, however in other circumstances, such as rural Mozambique, off-farm income has no discernible influence on calorie availability (Lencucha, Pal, Appau, Thow, and Drope, 2020).

One explanation could be that off-farm activities are generally available to men, whilst women are primarily responsible for food production. Non-farm income-generating activities are also vital for providing funds to pay for hired agricultural labor; women will have to rely more on inter- and intra-household links to provide extra labor. Cash revenue is also necessary for the purchase of planting materials, fish seed, and farm supplies because formal credit from banks is not accessible. Off-farm earnings, as well as cash and food crop sales, contribute to the creation of food and non-food markets with effective demand. With the monetization of the economy, non-market relations, such as informal or exchange labor, are regressing, while employed labor paid in cash and kind is expanding. Traditional gender roles in agriculture are changing in Nigeria, for example, resulting in increased participation of women in agricultural output due to rising male participation in non-farm activities and wage jobs (Kinkingninhoun, Komatsu, Mujawamariya, and Saito, 2020). Alternative productive activity with little capital requirements will be prioritized. These provide a quick return on investment and can include marginal workers. Any move into fish farming will be supported by an assessment of the required conditions, such as suitable land, availability of water and inputs, expected returns from available resources, and whether these are greater than those generated by other uses of the same resources, and expected market demand.

A homestead pond (Figure 1.2) serves multiple purposes and contributes to overall farm productivity. Fish is an important ingredient in human nutrition, serving as a source of animal protein and a prestigious food. Fish production has proven to provide excellent returns on land and labor, making it a profitable production; and fish is a high-value commodity. On the negative side, there are various barriers to fish farming acceptance that must be considered. These include the ability to master new technology know-how, the security of land tenure, which justifies the establishment of an expensive investment for the construction of the pond and access to fish seed and a market, which allows the fish to be sold. Because wetland lands are a common property resource in some nations, such as Malawi, access to them can be more challenging. The tenure arrangements are given meaning by social and cultural institutions that convince individuals that they are part of a secure, equitable, and well-adapted set of rights and duties. Also, if the investment payback period is short, land tenure security is less crucial. Fish farming in earthen ponds, irrigated rice fields, dams, and reservoirs, and pens or cages are the methods of fish farming available to farmers in rural or urban settings in developing countries.

Figure 1.2. Homestead pond.

Source: https://arkshore.org/wp-content/uploads/2020/02/Homestead-Fish-Pond-An-Extra-Source-of-Income.jpg.

The ponds are either drainable or non-drainable. Rice fields that have already been watered can be used. Stocking fish in dams, reservoirs, and natural water bodies can help increase fish productivity. Fish can also be reared in pens and cages in dams, reservoirs, and lakes. Freshwater aquaculture resources, which are dispersed over the length and width of India, have enormous potential to supplement fish supply. Within 5 to 6 years, substantial increases in annual fish production from the aquaculture sector could be achieved with concerted efforts to mobilize farmers to adopt fish farming Inland aquaculture production is currently expected to be over 3.0 million tons (Tomasso, 1997). Fresh water aquaculture has become a fast-growing farming activity in the country during the last two decades due to its flexibility in terms of operating area and scale of intensity, as well as its compatibility with other farming systems. Indian major carps or IMC and exotic carps like silver carp, grass carp, and common carp are the backbone of Indian freshwater aquaculture, accounting for more than 90% of total production (De Silva et al., 2006). Apart than IMC and foreign carps, there are various other fast-growing species that can be grown, such as minor carps like *Labeo calbasu* (Figure 1.3), *L. bata*, catfishes like singhi, magur, and murrels like Channa species, which are in high demand in specific parts of the country.

Figure 1.3. Labeo calbasu.

Source: https://d3i71xaburhd42.cloudfront.net/fdb639d2eaad10c69dd3e46152 6d4a9ae66862d2/2-Figure1-1.png.

1.3. AQUACULTURE

In India, there are roughly 2.41 million hectares of freshwater aquaculture land, the majority of which is under tropical warm water conditions and capable of growing production and productivity to about 5 tons per hectare per year (Figure 1.4). It may be conceivable to bring in roughly 8 lakh hectares of intense aquaculture in 5–6 years with better inputs in terms of seeds, feed, and fertilizers.

Figure 1.4. Aquaculture in India.

Source: https://faunalytics.org/wp-content/uploads/2021/03/fish-farm-india.jpg.

Another 50,000 hectares of new land can be brought into intensive aquaculture by establishing suitable forward and backward linkages ("Inland Fisheries | Department of Fisheries, GoI," n.d.). Such connections will necessitate the availability of high-quality fish seed, as well as the capacity building of farmers and businesses to adopt scientific fish farming practices and contemporary and efficient cold-chain systems to market fish in safe and hygienic circumstances. The current intensive aquaculture in ponds and tanks standards span a wide range of activities, from fish seed production to human resource development. Freshwater aquaculture development in the country offers a lot of promise. This potential has remained untapped for a variety of reasons. Apart from increasing production and productivity from current tanks and ponds, the NFDB plans to expand freshwater aquaculture to new locations, which would not only increase the country's fish production but also create new jobs, particularly in rural and interior areas. While fish seed is readily available in some regions of the country, farmers in other areas struggle to obtain high-quality seed in sufficient quantities. Further, the seed in such deficient areas has to be carried over great distances, which adds to the expense of inputs. To satisfy the needs of farmers in seed-deficit and new aquaculture areas, it is intended to assist entrepreneurs/farmers in the establishment of more hatcheries for the production of high-quality fish seed.

1.4. RESERVOIR DEVELOPMENT

Reservoir development requires high-quality fingerlings with a diameter of at least 100 mm (Figure 1.5). If the reservoirs are stocked with fingerlings measuring 100 mm and up, the yield will be higher (Rao, Imelda, Philipose, and Mojjada, 2013). Even though there is a demand for stock size fingerlings, there are currently insufficient seed growing facilities for rearing spawn or fry to fingerlings. As a result, infrastructure facilities for the growing of spawn or fry to fingerlings are essential. As a result, the scarcity of stock size fingerlings of cultivable freshwater fish species for stocking in reservoirs has been a limiting factor in the reservoir fishery production. In some sections of the country, the availability of fish fingerlings is adequate, but farmers in other places have difficulty obtaining high-quality fingerlings in sufficient quantities. Furthermore, in such deficient places, fish fingerlings must be delivered over a great distance, increasing input costs. As a result, it is proposed to assist entrepreneurs or farmers in establishing fish seed rearing units for the production of high-quality fish fingerlings in order to meet the need for fish fingerlings for supplementary stocking in reservoirs and ponds/tanks. Traditional pond culture methods are used in much of the world's aquaculture production. During the culture period, there is no water exchange in these ponds. To compensate for evaporation, some topping up may be necessary. Because of the difficulties in preserving water quality when there is a considerable biomass of cultured fish per unit volume of static water, static pond culture is usually extensive. To maintain productivity, rising biomass necessitates increasing fertilizer and supplemental feed inputs. This necessitates treatment of water quality issues such as unacceptable nitrogen compound levels and low DO levels at night. Supplemental aeration may make it possible to maintain DO while increasing biomass and production. Aerators, on the other hand, are frequently unavailable or impractical in rural areas where static pond culture is practiced.

Figure 1.5. Reservoir development.

Source: https://www.mdpi.com/water/water-12-02787/article_deploy/html/images/water-12-02787-g001.png.

The utilization of the environment as a fish farm is an open system, in which cultural organisms are confined or protected within the farm in a large body of water, with natural flows and maintaining water quality. There is no artificial water circulation in or out of the system. When a cage system is put in a big body of water, such as an ocean or an estuary, it is described as an open system. The fish in these cage systems are normally kept at a high density, and artificial feed is provided. Natural currents and tides, on the other hand, keep the water quality in check. As a result, these are highly open systems. Because there is no need for pumping, open systems have cheap operational expenses. The cost of capital varies tremendously depending on the culture. The most significant disadvantage of an open system is the huge variance in growth rate caused by seasonal variations in the environment. Water is added to compensate for evaporation or other unavoidable losses, or to preserve water quality. In most recirculating tank systems with vigorous cultivation, some water is discharged and refilled every day. This is due to features of the routine maintenance system, such as the removal of accumulated particles from filters. Water quality is far more difficult to maintain in entirely closed tank systems with intensive culture than in systems with a daily replacement rate of 5% or higher. Water quality in a recirculating tank system can only be maintained artificially, even with some limited water exchange each day. The high cost of building and manufacturing in intensive recirculating tank systems has hampered their commercial development as grow-out systems. The promise of high yields with year-round production close to markets, on the other hand, encourages their growth.

1.5. WATER QUALITY

Water quality (Figure 1.6) is influenced by feed supply, animal metabolism, wasted feed, and excrement output. To avoid toxicity, fish wastes like feces, uneaten feed, and ammonia must be constantly removed from the tank's water. The technique of solid waste removal is determined by the type of garbage. Dense solids sink and can be removed either continuously by gravity and flow or on a regular basis by siphoning. Alternatively, with continual agitation from aeration, water flow, and animal movement, settled materials can be kept suspended and removed using a mesh screen or allowed to flow into settlement basins or tanks.

Figure 1.6. Water quality.

Source: https://visipraxis.com/wp-content/uploads/2018/01/Blog-Aqua-Culture.jpg.

A swirl separator or hydrocyclone can also be used to extract settleable solids using centrifugal force. Air stripping, ion exchange, biological filtration, and algae removal are some of the technologies available for eliminating ammonia from water. Prior to air stripping, the water pH must be adjusted to 10 and then readjusted to culture levels near 7 before water can be returned to the rearing tanks (Stirling, 1986). The technology of ion exchange is expensive. Because of these factors, biological filtration remains the most used approach. Trickle filters, submerged filters, revolving biological contactors, packed tower filters, and fluidized bed filters are all examples of biological filters. Biological filters are designed to maximize the amount of surface area available for denitrifying bacterial adhesion. These bacteria, primarily Nitrosomonas and Nitrobacter species that can grow on practically any coarse media and are found in soil and water, are easily introduced into biofilters. First, ammonia is converted to nitrite, which is then further oxidized to nitrate. Because nitrite is hazardous to fish at low concentrations, both reactions must take place.

When oxygen consumption exceeds the capability of aeration using air, oxygenation is utilized to transport pure oxygen to water. Commercial sources of compressed oxygen gas and liquid oxygen are available, or oxygen generators can create it on-site. To create oxygen on site, the cost and stability of the electrical supply are usually the deciding factors. Because diffusers are ineffective at transferring gas to water, specialist components to provide oxygen to recirculating tank systems have been created. These components are capable of transferring more than 90% of oxygen to water. There are numerous oxygenation devices available (Espinal and Matulić, 2019). In recirculating tank systems, ultraviolet (UV) irradiation and ozonation are two means of continual disinfection. Larger-scale recirculation systems for pond culture are already being explored. It is difficult to function in ponds with intensive culture, such as shrimp culture, without regular water exchange. However, pond systems with extremely limited or no water exchange throughout the growth season are now being developed.

1.6. INTENSIVE FISH FARMING

The varying densities of organisms per unit volume or area are referred to as aquaculture intensity. It's useful for comparing the cultural levels of different species or similar species (Figure 1.7). However, when comparing densities of species from different groups, it is not possible. For example, intensive culture of tilapia in a recirculating system at 100 kg/m^3 of water is termed intensive culture; intensive culture of shrimp in ponds at 50 individuals per m^2 (1–2 kg/m^3) is considered intensive culture (Ray, 2012). The inputs into the system will be considered in order to sustain proper growth of the cultured organisms. It's called intensity because the higher the intensity of cultured organisms, the higher the demand for inputs into the system. Intensification aims to increase yield while using the same or less land, water, and labor inputs. Extensive aquaculture is distinct in that it is largely integrated into a natural ecosystem, relying on it for water quality and the majority of the animal's food and other needs. As a result, an extensive aquaculture system has few inputs to support fish growth and survival, such as basic organic fertilizers but no aeration. The natural productivity of feed inside the system, as well as natural gas exchange, are sufficient to support the cultured organisms in these systems, which typically have an uncertain stocking density. Low-intensity aquaculture produces only a minor improvement in productivity over natural levels.

Figure 1.7. Intensive fish farming.

Source: https://www.openaccessgovernment.org/wp-content/uploads/2021/03/dreamstime_l_79602113-scaled.jpg.

High water exchange rates and, in some situations, mechanical measures are used to maintain water quality. In intense culture in indoor tanks, particulate waste collection, gas exchange and oxygen production are all undertaken by mechanical means. Particulate waste settles, bacteria decompose it, and mechanical aeration improves gas exchange in outdoor intensive systems with a soil substrate and phytoplankton. In intensive systems, stocking density varies widely depending on the system and the cultivated organism, although it is always quite high. In addition to the natural food in the pond, supplemental feed must be added. Tilapia culture is particularly well suited to developing countries because it offers a wide range of management and capital investment options. Management tactics in lower stages of intensification entails the use of fertilizers to boost natural productivity and to improve the levels of dissolved oxygen (DO). The stocking rate is between 5 and 10 fish per m^3. Such techniques have been found to produce higher fish yields than natural unfertilized systems (Noltan, Tharwat, and Samra, 2005). Monoculture is the only technique of culture employed in running water systems, recirculation systems, and cages where natural food supplies are restricted. Because the natural food available in the water column is not fully absorbed by fish in pond culture with fertilization monoculture of single fish species, a combination of fish species with different feeding patterns is suggested.

1.7. FISH PRODUCTION

The possibility of enhancing fish production by better utilizing natural feeds is the most essential concern in polyculture. Species successfully stocked together differ in their feeding habits and occupy different trophic niches in the pond. The feeding habits of Tilapia species, common carp, silver carp, and gray mullet differ. In Egypt, common carp polyculture with all-male tilapia, and occasionally silver carp, is a frequent practice. Tilapias and silver carp both graze on algae, assisting in the maintenance of a balanced biological environment free of algal blooms.

Polyculture of all-male tilapia with marine or fresh-water prawns produces a more balanced biological habitat in ponds than prawn monoculture alone (Figure 1.8). As in the case of silver carp and tilapia, the pond oxygen regime improves. The consumption of surplus algae by silver carp enhances the balance of oxygen production and consumption. Tilapia eat the pond's organic matter (OM), which increases oxygen consumption when bacteria digest it. As a result, tilapia enhances pond oxygen levels. Some fish, such as tilapia, eat the excreta of other fish, such as common carp, which eat the excreta of silver carp. Because the latter do not digest all of the algae they eat, the larger algae become available to common carp, which are unable to consume all of the free algae in the water. A system of producing fish in conjunction with other agricultural and livestock farming operations focused around the fish pond is known as integrated fish farming. The farming sub-systems, such as fish, crop, and livestock, are linked in such a way that by-products or waste from one sub-system become valuable inputs to another, ensuring total utilization of the farm's land and water resources and hence maximum farm production with minimal financial and labor expenditures. Rice fish farming is a good farming technique. This fish grows in the thick rice plants, hidden from birds and other predators, while also providing fertilizer with their compost, eating insects and pests that are harmful to rice plants, and assisting the rice field in oxygen circulation.

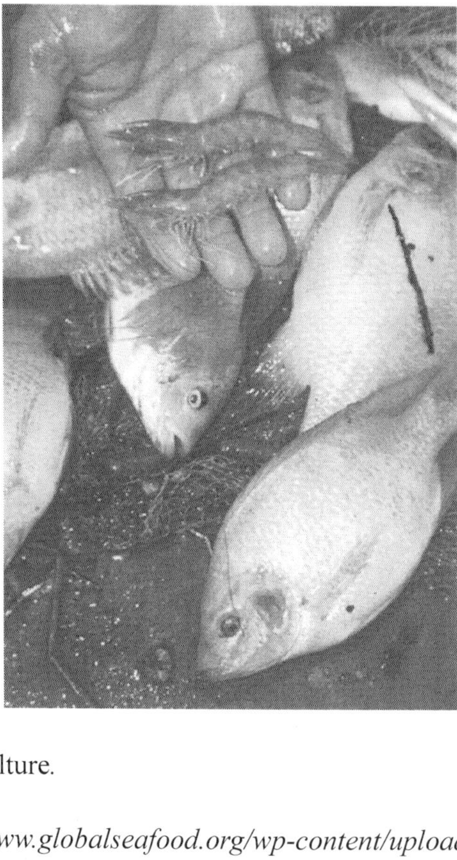

Figure 1.8. Polyculture.

Source: https://www.globalseafood.org/wp-content/uploads/2001/06/Fitzsimmonspic1.jpg.

Rice and fish farming (Figure 1.9) can boost rice yield by 10% while also providing an additional food source. China is the largest producer of rice-fish culture, with an area of roughly 1.3 million hectares of rice fields with various forms of fish culture, producing 1.2 million tons of fish and other aquatic animals in 2010 (Hu et al., 2021).

Figure 1.9. Rice and fish farming.

Source: https://www.agrifarming.in/wp-content/uploads/2016/10/Rice-and-Fish-Farming.jpg.

1.8. CHALLENGES

The challenges that rice–fish farming faces are similar to those that face general aquaculture development. They include seed, feed, and capital availability and access, as well as natural risks such as water control, disease, and predation. Freshwater is becoming one of the scarce natural resources, and competition for it is one of the most pressing issues confronting emerging countries. Water is a valuable resource in rice–fish farming because it boosts productivity per unit of water consumed. Profits vary based on production features, but income increases of up to 400% have been reported when compared to rice monoculture, and they might be considerably higher if high-value aquatic species are farmed (Freed et al., 2020). To begin, a farmer must take practical steps to determine an ideal site.

They must locate a location that is not flood-prone in order to prevent your rice fish farming from being washed out. Make a 60-cm-high dike around the field's perimeter and this method serves two purposes. It keeps farmed fish in the field while also allowing for vegetable cultivation around the rice field. The next phase is to dig a channel to create a safe haven for fish during the dry season. Aquaponics is a technique that combines aquaculture that produces aquatic animals such as fish, crayfish, or prawns in tanks and hydroponics that involves growing plants in containers. The excretions of the animals being reared might accumulate in the water in typical aquaculture, causing toxicity. The aquaculture element of aquaponics is used to raise aquatic animals, while the hydroponics half is used to cultivate plants. In high proportions, the effluent-rich water is hazardous to aquatic animals, yet it includes nutrients necessary for plant growth. Aquaponics systems are frequently divided into many components or subsystems that are responsible for the successful removal of solid wastes, the addition of bases to neutralize acids, or the maintenance of water oxygenation (Thorarinsdottir, 2015).

1.9. HYDROPONICS

A deep-water culture hydroponics system (Figure 1.10) is a system where plants grow directly into the effluent rich water without a soil media. Many plants are suited for aquaponic systems, but which one functions best for a given system is determined by the fish's maturity and stocking density. These variables determine the quantity of nutrients in fish effluent and how much of that nutrition is made available to plant roots via microorganisms. Green leafy vegetables with low to medium nutrient requirements, such as tomatoes, cucumbers, and peppers, are well adapted to aquaponic systems. Freshwater fish are the most common aquatic animals cultivated in aquaponics, however freshwater crayfish and prawns are also utilized occasionally. Saltwater aquaponics is a type of aquaponics that uses saltwater fish. Warm-water and cold-water fish of various species adapt well to aquaculture systems. Because tilapia is a warm-water fish that can handle crowding and changing water conditions, it is the most preferred fish for home and commercial initiatives aimed at raising edible fish. Bluegill and catfish are ideal fish species for home systems in temperate locations where the ability or willingness to manage water temperature is lacking (Ficke, Myrick, and Hansen, 2007). If the fish in the system do not need to be edible, koi, and goldfish can be employed. Channel catfish, rainbow trout, perch, common carp, and striped bass are all good choices.

Figure 1.10. Hydroponics system.

Source: https://iberflora.feriavalencia.com/wp-content/uploads/2018/04/CDC_South_Aquaponics_Raft_Tank_1_2010-07-17-1024x768.jpg.

1.10. NITRIFICATION

Nitrification (Figure 1.11) is one of the most significant activities in an aquaponics system since it minimizes the toxicity of the water for the fish while also allowing the plants to remove the resulting nitrate compounds for sustenance. Ammonia is continuously discharged into the water as a by-product of fish metabolism through excrement and gills, but it must be filtered out of the water since greater quantities of ammonia will kill fish. The bacterium is responsible for converting ammonia to usable nitrates for plants in an aquaponics system.

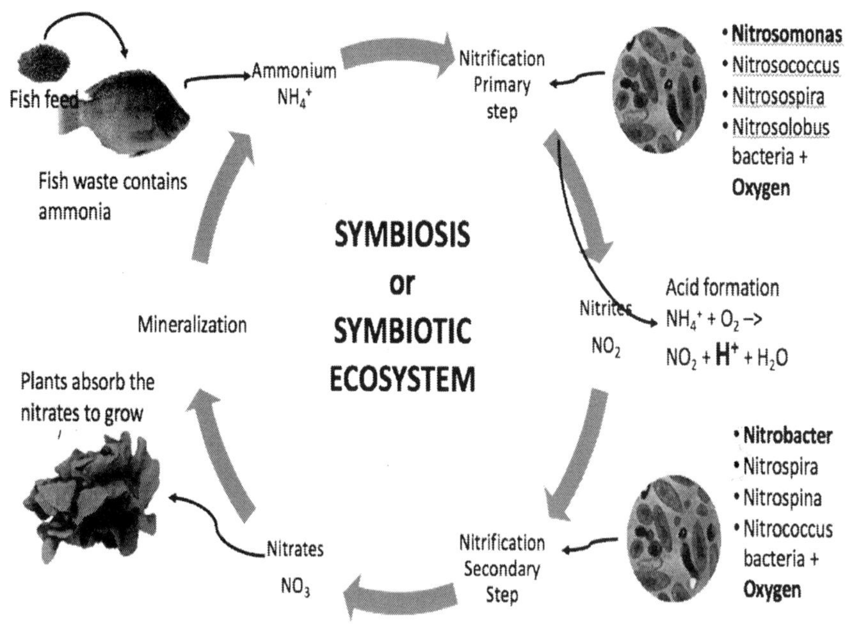

Figure 1.11. Nitrification.

Source: https://urbanspaceaquaponics.com/wp-content/uploads/2018/01/Symbiosis_and_nitrification.png.

The combined surface area of the submerged roots of the vegetables is large enough for numerous germs to accumulate. The speed of nitrification is determined by the surface area, as well as the quantities of ammonia and nitrites in the water. The full assimilation of ammonia and nitrite requires the proper care of these bacterial colonies. This is why most aquaponics systems come with a biofiltering unit to aid in the growth of these bacteria. In normal operation, aquaponic systems do not discharge or exchange water, but instead recirculate and reuse it very effectively. To maintain a stable aquatic environment with minimal fluctuations in ambient nutrient and oxygen levels, the system depends on the relationship between the animals and the plants. Water is only given to replace water lost due to plant absorption, surface water evaporation, and biomass removal from the system, such as settled solid wastes. As a result, aquaponics consumes around 2% of the water that a traditionally irrigated farm would use to produce the same amount of vegetables. This enables aquaponic agricultural and fish cultivation in regions where water and fertile soil are scarce (Joyce, Goddek, Kotzen, and Wuertz, 2019).

1.11. GROWTH

Low stocking densities are utilized in widespread aquaculture, and no extra feed is offered. Fertilization can be used to encourage the growth and production of natural food in the water. Carp culture does not necessitate water replacement throughout the culture period, but shrimp culture does, which is accomplished through tidal flushing, in which new water is introduced during high tide and used water is drained during low tide. Large ponds are typically utilized for substantial aquaculture. Shrimps produce less than 0.5 ton per hectare per year, but carps produce 1 to 3 ton per hectare per year. Ponds in this system range in size from 1.0 to 2.0 hectares. Stocking densities range from 30,000 to 40,000 shrimp PL per hectare to 5,000 to 7,500 carp fingerlings per hectare (Ali et al., 2022). In this approach, organic and inorganic fertilizers are used to fertilize the ponds, and phase manuring is done once a month. Low to moderate amounts of supplemental feed are given. In carp culture, water is replaced every two to three months at a rate of 10 to 20%, while in shrimp culture, water is replaced every two weeks at a rate of 10 to 20% from the second month onwards. In shrimp cultivation, productivity averages approximately 1.0 ton per hectare, while carp culture produces 4 to 5 tons per hectare (Das et al., 2004). Semi-intensive aquaculture uses 0.5-hectare ponds with higher stocking densities than extensive aquaculture. Supplemental feeding occurs in little amounts. Water replacement is done once or twice a month at a rate of 10% in carp culture, and once or twice a week at a rate of 10% to 20% in shrimp cultivation. Shrimp culture produces roughly 1.5 tons per hectare, while carp culture produces 5 to 7 tons per hectare (Nisar, Peng, Mu, and Sun, 2022).

In intense aquaculture, the pond size is typically small, around 0.2 hectare, for operational convenience. Very high densities of culture organisms, like 300,000 to 500,000 shrimp PL/ha or 20,000 to 25,000 carp fingerlings/ha, are stocked in intensive aquaculture (Harkes et al., 2015). The system is completely reliant on formulated feeding. The stock is fed at predetermined times. The calculated daily feed ration is provided in equal amounts from three to six times a day in intensive shrimp culture. Water replacement is typically 25% to 30% on a daily basis in an intense culture. Production is substantially higher in intensive aquaculture, for example, 8 to 10 tons per hectare per crop in shrimp culture and 12 to 15 tons per hectare in carp culture (Hortle, 2009). Semi-intensive and intensive aquaculture operations require more labor and are more expensive to start up and manage than extensive aquaculture systems. Apart from that, they have a higher chance of death due to disease outbreaks, poor management, and oxygen depletion due to aerator failure during power outages. However, when compared to the extensive level, the yield rate and financial returns are substantially more appealing at the intense and semi-intensive levels. However, due to the high cost of inputs used in intensive aquaculture, studies have shown that the returns on investment from semi-intensive aquaculture are better than from intense aquaculture. Intensive aquaculture operations are appropriate for countries with limited land and water resources, whereas extensive aquaculture is ideal for the opposite scenario. Intensive aquaculture is widely used in countries such as Israel and Japan (Sumuduni, Munasinghe, Chandrarathna, and Amarasinghe, 2015).

Intensive fish farming has become the primary source of marine farmed products in less than 30 years (Naylor et al., 2021). Nonetheless, extensive and semi-intensive systems produce enormous amounts of output and occupy large areas along the southern European coasts. Coastal lagoons, natural and managed deltas, and semi-closed bays and estuaries, as well as earthen ponds. The total surface area of coastal wetlands employed for substantial and semi-intensive cultivation today appears to be rather large. For modern semi-intensive farming, productivity is a concern, but good results can be achieved. This is not the case with large systems, which display all ranks of values based on natural restrictions, management, and technical capabilities, among other things. In many circumstances, productivity gains are possible. Lower values are frequently determined by the extent of manager participation and whether or not an economic return is feasible. Farmers have incorporated modern techniques into the bulk production of shellfish, as well as semi-intensive fish and shrimp husbandry. They are pretty similar from one country to the next. Traditional traditions, on the other hand, show considerable disparities in terms of technical standards and water management. Improvements in frequently farmed species have enabled more consistent output, particularly in monocultures of oysters and mussels. Shellfish aquaculture produces huge quantities of shellfish in lagoons, making a significant contribution to total European shellfish production. The utilization of lagoons has been optimized, but pollution and eutrophication processes have hampered the expansion of many other lagoons. Because of the traditional background of local communities, historical traditions of massive fish farming are still practiced. They create tiny quantities of high-quality goods but lack traceability and public awareness. Only Italy, with its various types of lagoons, makes optimal use of its potential for broad fish production, with likely huge production volumes despite the lack of statistics. Semi-intensive fish production is a natural progression from these traditional methods. Several successful trials conducted over the last 15 years have since been transformed into viable farming systems (Meemken and Qaim, 2018).

1.12. FACTORS

Several factors have led to the poor growth rate of Portuguese aquaculture. Traditional extensive systems and newer semi-intensive systems have been established in wetland areas, some of which are protected natural parks. This regulatory restriction is a major roadblock for the sector, particularly for semi-intensive farms, as it prevents new coastal farms from being built and current ones from being expanded. Due to limited growing opportunities, the majority of farms are micro-enterprises, producing tiny quantities and selling to local buyers. In addition, the vulnerability of coastal ecosystems to untreated anthropogenic wastes from urban and agricultural sources has resulted in significant economic losses, particularly in mollusks production in lagoon systems, due to weak coastal management strategies. Significant-scale fish farming in earthen ponds is experiencing difficulties and is increasingly disappearing, resulting in the abandonment of large regions along the Portuguese coast. Because of their low profitability and lack of interest among farmers, current research is focused on semi-intensive and intense production difficulties. Due to the rising density of production, efforts have been made to detect and prevent early indicators of disease. Farmers have also devised methods to reduce losses from predators such as otters and birds, which can amount to up to 30% of total yield. Intertidal aquaculture operations are still highly reliant on environmental limits and nearby activity like agricultural, industry, aquatic sports, or even other fish farms (Curtis, Pitt, and Conover, n.d.). The decline in output is mostly due to a parasite called *Perkinsus marinus* that causes high mortalities and enormous production losses, but it is also related to challenges in controlling predators and an excess of OM accumulation in the nursery sediments. Long-line aquaculture has helped to solve some of these issues. Aquaculture development requires efforts in numerous areas, including production and management, marketing, research, and training. Instead of focusing on the same common species, diversification would be an issue. In this sector, research and experimentation are critical for innovation, as is a wider interaction between industry and other stakeholders. Farmers' associations, for example, could provide better training to their members.

Despite a number of challenges, such as administration and other sectors of society's lack of understanding, and rivalry from other economic activities, the use of salt ponds in modified marshlands for aquaculture persists. The majority of these coastal marshes and lagoons have been designated as natural parks or protected areas because of their great environmental value. Although this designation prevents alternative uses that are incompatible with the preservation of wetlands, it does not guarantee the preservation and health of some of these ecosystems. The ponds tend to fill with mud or be destroyed by the sea unless farmers make massive and constant efforts to control the water and ditches. Many examples of these repercussions can be found at the abandoned salt works of Cádiz Bay. Those zones, on the other hand, have a lot of promise for coastal aquaculture. There is a growing interest in building or using new earthen ponds, which is higher than the number of new requests for marine cages. This fact suggests that marine fish productivity in earthen ponds is likely to improve in the near future. However, in many situations, the essential investments are contingent on administration approval and/or the resolution of land property disputes.

Several issues are currently plaguing the vast French extensive and semi-intensive aquaculture business. Shellfish production, particularly oysters, suffered from severe diseases in 2008 and 2009, and natural reproductive success has been low. Unbalanced coastal conditions are thought to be aided by climatic events, which help toxic microalgae, bacteria, and viruses generate fatal outbreaks (Hinder et al., 2011). Higher sensitivity of cultivated species to worse water quality could possibly be due to urban pressure and continental runoff. Large-scale water purification operations at the river catchment scale are finally showing results, and will hopefully compensate for the rise in human population on the coast. Farmers and scientists have devised engineering solutions to address technical issues such as water availability, water quality, farming process optimization, and sanitary insurance. Despite rising consumer demand, the utilization of marshlands remains steady. In coastal wetlands, there is plenty of room for oyster storage and polishing. Lagoon productions are also of great significance. Water quality will continue to be a major limiting issue unless regional water management can achieve significant, large-scale improvements. Shrimp production is subject to micro-market conditions, but because of its high market position, it is still in good shape. However, increasing environmental quality limits or a decline in larval quality could eventually drive-up production costs.

Private landowners' interest in these productions has been declining for the past 50 years, and optimization of production procedures has been advocated to encourage them. Semi-intensive fish farms are still productive, although their spatial significance in marshes is limited. Their numbers have remained stable due to administrative restrictions. A few companies use their facilities to house nurseries and supply fry to local farmers. Because of its low environmental impact, this activity could become more important, but it still requires farmers to improve their skills, which would require additional training and courses for the majority of them. Valliculture has long been a good example of efficient use of coastal regions for productive purposes in the North Adriatic areas, and it has made a significant contribution to the conservation of wetlands in the setting of coastal lagoons (Cataudella, Crosetti, and Massa, 2015). The term valle, which means embankment, is presently only applied to coastal basins in the North Adriatic. It originally referred to a simple rudimentary lagoon that was fenced for fish farming. For many birds, Valli provides an ideal setting for nesting, wintering, and feeding supplies. Valli systems are low-energy farming approaches that have helped to protect sand banks and shallow lagoon waters. Increased eutrophication caused by aquaculture and agricultural operations, as well as the discharge of treated and untreated municipal wastewater, has gradually altered the environment in some lagoons.

The use of valli as seeding and growing regions for aquatic species is widely acknowledged as an environmentally friendly exploitation strategy, but it must be managed and controlled to preserve exploited fish stocks while minimizing negative impacts on natural fish resources. For this goal, a set of recommendations for fish stock management has been produced. It includes monitoring during the farming cycle, management of water level, hydrodynamics, and fresh and salt water stratification to avoid extreme temperature effects on fish survival, anticipation of recapture time from November to September to avoid current abundance of fish-eating birds and migratory fish aggregations, and optimization of seeding period and quantity of autochthonous species. The key bottleneck in valliculture is the increase in fish-eating bird populations, which has resulted in a sharp fall in recapture rates during the previous 10 years. Their impact on stocked fish could be direct, such as when they catch them, or indirect, such as the development of secondary infections in escaped fish as a result of bird injuries, or a general decrease in performance due to various factors like stress, lateral line anomalies as a result of scars, lower accessibility to local sites with higher trophic availability where predator bird presence is higher. Furthermore, they disrupt fish movement toward fish barriers, dispersing fish toward summer pasture zones and causing cold-related death throughout the winter.

Valliculture is an effective paradigm for long-term coastal wetland management in areas with varying salinity. There are some areas in Greece where extensive coastal aquaculture can be carried out. In terms of production volume, however, most aquaculture development efforts are focused on intense production in sea cages. Nonetheless, existing comprehensive systems could be improved in a number of ways. Maintenance and upgrading of entering channels, as well as fish facilities, are among the important efforts for increasing estuary production. When macro-algal blooms occur, poor water quality threatens extensive fish productivity in lagoons. Technical personnel are often in charge of removing overgrown vegetation. For semi-intensive production, efforts are concentrated on preventing theft in the lagoon area and building ditches, which are especially crucial in earthen ponds in Northern Greece. The depth of the ditch is critical to the effectiveness of this operation. In addition, depending on the stocking density of the ponds, aeration of the lagoon and ponds with aeration paddle-wheels may be required throughout the summer. Predator fish such as sea bass must be removed in order to avoid catastrophic losses in stocked populations. Furthermore, environmental monitoring is necessary to ensure the viability of fish populations and the safety of the product for consumers. The majority of locations used for extensive and semi-intensive aquaculture are in areas that receive agricultural and rural effluent, which may contain pesticides and other hazardous substances. Lagoon and pond management thus necessitates both technical and scientific skills. Experienced employees or support from public services with the relevant scientific and technical background can provide this expertise. The majority of fish farmers' efforts are focused on maintaining fish traps, removing plant detritus, and sorting fish.

1.13. LEGISLATION

Adequate legislation and governance implementations promoting substantial aquaculture in natural wetlands and the ecological management of their functionality and productivity should aim for a best practices formulation, taking into account the challenges. In order for farmers to maintain a large-scale operation while earning enough money, research would be beneficial in boosting natural productivity to assure enhanced basic products. Engineering enhancements in semi-intensive systems would also assist rearing processes by increasing dependability and security. Farmers who still rely on observation clues alone to control production should be provided with water quality survey expertise, rather than mitigating problems using fundamental survey concepts like DO, salinity, temperature. Land production potential is based on wetlands morphology management and restoration. Due to a lack of hydraulic intervention in many locations, this potential is dwindling by the day, and the capacity to reuse such hydro systems would incur additional expenses to restore proper design and hydrological regimes. A legislative framework should be developed to enable for the quick evaluation, regulation, and implementation of hydraulic and other works in coastal lagoons. Such projects should be done in collaboration with farmers and government organizations whenever possible.

The ecological functionality of vulnerable ecosystems like lagoons, marshes, and wetlands is tightly tied to specialized management practices and planning, which are currently mostly supported by private corporations. The identification of relevant multilayer ecological descriptors, as well as their continual refinement, selection, and monitoring, is required for the management of these ecosystems. Integrating and elaborating existing data should be utilized to evaluate ecosystem dynamics, structure, and function in order to forecast the direction in which the ecosystem under human control is evolving and hence manage it with suitable interventions. One of the few viable pathways for maintaining the profitability of this ecologically friendly animal production and countering market saturation of species such as sea bream and sea bass generated by intensive fish breeding is to identify the products of extensive aquaculture. The ability to combine aquaculture with multifunctional, i.e., tourism, educational, recreational activities involving new generations of managers is a key issue for management continuity and traditional aquaculture conservation, both of which are critical for ecosystem protection. However, such activities may not be enough to ensure that extensive aquaculture is economically profitable. The extensive aquaculture sector's development policies should be equipped, as was done for rural development in agriculture, with instruments to support enterprises and designed to reward the use of areas with high environmental value methods.

CHAPTER 2

INTENSIVE FISH FARMING

CONTENTS

2.1. Introduction..28
2.2. Culture Systems..30
2.3. Foods..31
2.4. Intensive Systems..32
2.5. Problems..34
2.6. Aquaculture Sector..35
2.7. Nutritional Requirements..37
2.8. Aquatic Environment..38
2.9. Fish Appetite..41
2.10. Receptors..43
2.11. Freshwater Fish..46
2.12. Analysis..50

2.1. INTRODUCTION

It is generally known that formulated aquaculture feeds are successful in intensive aquaculture. Modern aquaculture has benefited from a better understanding of the nutritional requirements of certain fish and shrimp species, as well as advancements in feed manufacturing technology and feeding practices (Hochheimer and Wheaton, 1997). While our understanding of the nutritional needs of most farmed species is still incomplete, a global aquafeeds business has developed and commercially produced feeds for a wide range of species. Many farmers also make their own feeds from locally available components, using formulas established and distributed by development agencies. Formulated feeds are used in the diets of farmed fish and shrimp to supplement or replace natural feeds. In global aquaculture, there is a general trend toward increasing the intensity of cultivation methods (Figure 2.1). This increase is mostly due to the broad availability of aquaculture foods, which has been accompanied by improved water quality management, disease control, and general husbandry techniques.

Figure 2.1. Intensive fish farming.

Source: http://fishcount.org.uk/wp-content/uploads/2009/06/edsalmoncages-norway.jpg.

Some culture techniques have been used to sustain intensive culture, such as Atlantic salmon farming. Feeds have been incorporated into conventional fish and shrimp culture methods in other parts of the world to accommodate more intensive culture techniques with higher production densities. This trend may be seen in the culture of giant tiger shrimp (Figure 2.2) in Southeast Asia, as well as tilapia culture throughout the tropics and subtropics.

Figure 2.2. Giant tiger shrimp.

Source: https://divcomplatform.s3.amazonaws.com/www.seafoodsource.com/images/d651ccb8695781c74e0480ca1085364c.jpeg.

2.2. CULTURE SYSTEMS

Although the comprehensive, semi-intensive, and intensive terms are frequently used to describe culture approaches, they are arbitrary terminology with indistinct boundaries in practice. They are, however, often tied to the amount of feed and/or fertilizer inputs, as well as the maximum fish and shrimp stocking densities that may be supported. The majority of aquaculture fish and shrimp are raised in extensive and semi-intensive pond farming systems. Organic and inorganic fertilizers are used in vast pond culture production. Pond fertilization encourages the growth of simple plants, which are the foundation of the pond food web.

Phytoplankton (Figure 2.3), bottom-dwelling crustaceans, and smaller fish are all consumed by fish or shrimp introduced in such ponds (Reaka-Kudla, 2001). The amount and type of fertilizers used by different management strategies varies greatly. Extensive aquaculture is most effective when it is combined with other types of crop and livestock production. Animal manure and agricultural by-products are employed as sources of carbon, nitrogen, and critical minerals in these systems to boost phytoplankton primary growth in the presence of sunlight. A variety of natural food species allows several distinct species of fish to be cultured in the same pond habitat, each feeding on a different organism. The ponds' production capacity is greatly increased by this polyculture approach. This sort of fish farming is most popular in tropical and subtropical freshwater ponds, where production sizes range from small families and communities to major commercial undertakings.

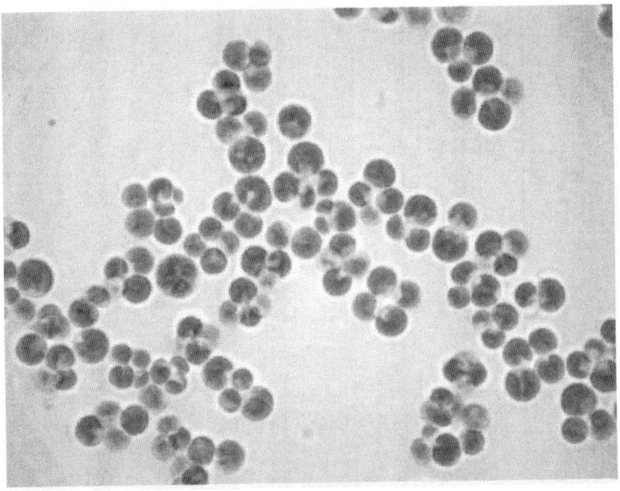

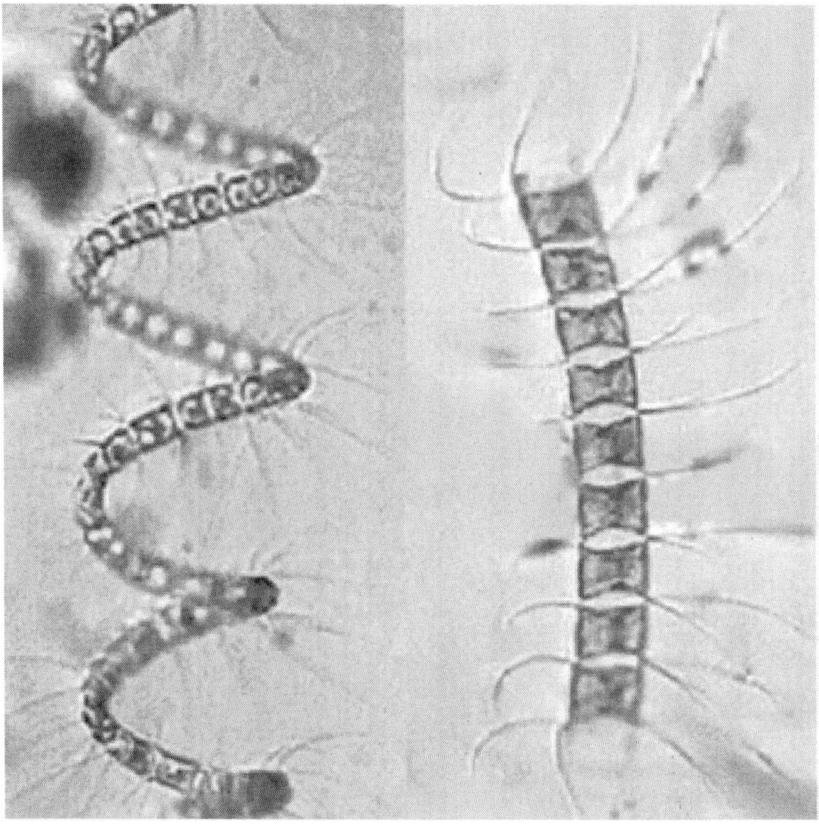

Figure 2.3. Phytoplankton.

Source: https://www.globalseafood.org/wp-content/uploads/2017/04/POH-Pic-1-960x368.jpg.

2.3. FOODS

Supplementary foods may be used to raise the production of fish from pond systems above the amount provided by the availability of natural food. Supplementary feeds range from cereals to agricultural and fishing by-products to designed diets, depending on the species of fish or shrimp being farmed. Supplementary feeds are typically made from locally available components and are either created by the farmer or obtained from local suppliers. Supplementary meals are frequently nutritionally deficient and would not suffice as a single source of sustenance. Their purpose is to supplement the basic nutrients obtained by the fish or shrimp from natural food items by providing additional major nutrients. In semi-intensive cultures, more expensive nutritional feeds are employed to complement natural food production when economic constraints allow. Silver carp, common carp, grass carp, and bighead carp are all widely cultivated in extensive or semi-intensive pond systems.

Tilapia culture methods are farmed in a variety of ways ranging from extensive to highly intense. The majority of species utilized in aquaculture are *Oreochromis niloticus*, *O. aureus, O. hornarum, O. mossambicus*, and *Tilapia zilli*, which are available as pure or hybrid strains (Bradbeer et al., 2019). These species graze on a variety of plankton, benthic invertebrates, plants, and algae in large culture systems. Early fish culturists identified these eating patterns as an important property of the group, and culture methods were initially focused on improving natural pond feed sources with lime and different fertilizers. In some instances, the usage of additional feeds to improve yields from specific pond areas has evolved naturally. Supplemental feeds are widely used in tilapia pond culture and rather sophisticated methods have been devised to boost yields. The parameters that influence productivity in these systems are complicated, and expert management approaches are required to balance various fertilizers and feed inputs while maintaining appropriate fish stocking rates. Tilapia intensive culture is less prevalent, although it accounts for a growing share of global production. Intensive cultivation is carried out in raceway systems, tanks, or floating cages or enclosures with flowing water. The fish have little or no access to natural food organisms in these environments, thus they must be fed nutritionally complete feed. High stocking densities are employed to optimize the farm's output potential, while water exchange rates are managed to maintain adequate levels of dissolved oxygen (DO) and dilute metabolic wastes.

2.4. INTENSIVE SYSTEMS

The levels of intensification used in aquaculture are linked to the farm's operational goals. Intensive systems are commonly employed to produce high-valued species, though not exclusively. This is required in order to justify their high capital and operating expenditures. Fish and shrimp farms are typically run to earn cash through national and worldwide sales. Extensive aquaculture systems, on the other hand, are typically created to suit local and regional fish supply needs. Semi-intensive farms can perform one or more of these functions.

Many places have a range of culture methods for the same species running concurrently. The origins of shrimp aquaculture (Figure 2.4) can be found in Southeast Asia's coastal brackish water ponds. The operation of tidal ponds or tambaks can be traced back to the 14th century in Indonesia. Post-larvae and juveniles enter the pond with the tidal input during migratory activities, where they are then held by mesh gates and screens in the traditional functioning of tambaks. The species eat natural algae, plankton, and a variety of benthic species that thrive in the warm shallow ponds, and are later harvested at the entrance when they try to make their way back to the ocean. Mixed shrimp population production in these systems is highly unpredictable, and it is frequently a supplementary crop to a main milkfish harvest. Under ideal conditions, production might reach 200–250 kg/ha, with two harvests per year possible (Yap, Villaluz, Soriano, and Santos, n.d.).

Figure 2.4. Shrimp aquaculture.

Source: https://static.toiimg.com/thumb/msid-70564848,imgsize-95106,width-400,resizemode-4/70564848.jpg.

The trend toward more intense shrimp aquaculture (Figure 2.4) is similar to that of finfish farming. More intense cultivation methods for a variety of shrimp species were created in response to scientific achievements in the domestication of the kuruma shrimp in Japan. Most farmed shrimp are currently grown in semi-intensive systems, where they are fed a combination of natural prey organisms and prepared diets. Penaeus monodon is the most frequently grown shrimp species, accounting for more than a third of all farmed shrimp worldwide. Production levels vary, but are usually between 1.5 and 2.0 tons per hectare per year. In shrimp aquaculture, intensive methods are employed to enable high density production through the use of formulated meals (Boyd et al., 2017). Based on two production cycles, these systems might produce up to 10 tons per year. In some areas, the expansion of shrimp farming methods has resulted in significant production losses due to pollution and disease transmission. In intensive shrimp production, poor feeding techniques can be a serious issue, causing pollution both within ponds and in nearby coastal waters, as well as promoting an environment where shrimp are stressed and opportunistic infections thrive.

2.5. PROBLEMS

Despite issues, intense culture systems are becoming more commonly used, and they now account for a growing part of the world's farmed shrimp. Over 90% of new farms erected between 1989 and 1991 in Thailand, one of the largest producers of farmed shrimp, were designed for intensive operations (Huitric, Folke, and Kautsky, 2002). Because natural food organisms play a substantial role in both systems, but the overall feeding dynamics are poorly understood, distinctions between semi-intensive and intensive production strategies are less obvious in shrimp farming than in fish farming. Experiments using stable carbon isotopes revealed that natural food organisms accounted for 60–70% of shrimp growth in semi-intensive systems, whereas formulated feeds accounted for the remaining 30–40% (Jatobá et al., 2014). Because there is no easy way to evaluate the percentage of natural foods in farmed shrimp diets, farmer conversion rates based on the quantity of formulated feeds utilized may exaggerate feed efficiency. More research is needed to assess the amounts of formulated feed uptake as well as the fate of uneaten food in the pond ecology.

2.6. AQUACULTURE SECTOR

The rise of the aquaculture feeds sector has been fueled by the introduction of formulated diets for trout and salmon. Our knowledge of these species' nutritional needs exceeds that of most other species, and manufacturers have developed an increasingly complex range of trout and salmon diets. Prior to the 1950s, replenishing trout and salmon were fed a variety of wet diets made from minced garbage fish at hatcheries across Europe and the United States. Freshwater rainbow trout farming was one of the earliest intense industries to emerge in Denmark. This pioneering sector, which began at the turn of the century and flourished significantly during the 1950s, was founded on the consumption of by-catch marine species. The farms' proximity to Danish North Sea fishing ports enabled daily supply of fresh feeds to the fields. The advent of the Oregon Moist Pellet in the early 1950s marked a big step forward in the development of formulated feeds. This diet comprised pasteurized wet fish products to reduce the danger of disease transmission, as well as vitamin and mineral supplements. The Oregon Moist Pellet is still commonly employed in freshwater hatcheries to produce Pacific salmon smolts. It is produced and sent frozen to hatcheries, where it is thawed before use. Researchers in Europe and the United States were working on dry feed formulas for trout at the same time as they were researching moist feeds for salmon hatcheries.

In the Southeastern United States, extensive channel catfish farming grew at the same rate as trout and salmon farming in Europe. The nutritional requirements of catfish were gradually recognized, and dry feed formulations were gradually adjusted in light of new findings, to the point where nutritionally complete, dry feeds can now be made economically for use in intensive culture. Basic nutrition research has aided the development of intense culture methods for the kuruma shrimp, *Penaeus japonicus*, in Japan. During the early stages of development in the 1960s, cultured shrimp were fed fresh food such as bivalves, squid, or scrap fish. As more information about the nutritional requirements of shrimp became accessible in the 1970s, these methods were followed by the gradual introduction of prepared feeds. These early advances have prompted research into the dietary requirements of a growing range of fish and shrimp species. While the nutritional principles that must be followed in feed development are the same for all animals, there are considerable variances between fish and shrimp species. All animals require about 40 basic dietary elements, including amino acids, fatty acids, vitamins, and minerals.

The majority of intensively farmed animals are carnivorous, with high protein and calorie requirements in their diets. This has prompted the creation of nutrient-dense diets in some areas of intensive aquaculture, as well as a reliance on high-cost, high-nutrient-quality sources like fish meals and marine oils. Some fish species' feed formulas have altered dramatically in recent years. These modifications demonstrate a considerable decrease in phosphorus incorporation and an increase in the usage of lipids to provide energy. The necessity to generate diets that are water stable complicates feed development for aquaculture species. This requirement is especially important for shrimp and other fish species that feed slowly over several hours. Diets must be effectively bound to avoid breakage in the water, and some nutrients must be leached as little as possible. Once aquaculture feeds are immersed in water, water soluble vitamins, free amino acids, and certain minerals are rapidly lost. These are issues unique to aquaculture feeds, necessitating a re-evaluation of ingredients, feed formulas, and manufacturing methods.

Vitamin C in protected forms with low solubility is now frequently used in aquaculture feeds. Aquaculture feeds differ from normal livestock feeds in a number of important physical and chemical features. Pellets of varied densities can be produced using certain feed manufacturing technologies. As a result, fish farmers can choose pellets that float or sink at varying rates and are suitable for use with specific species in specific types of rearing systems. Because shrimp are bottom feeders, shrimp farmers prefer fast sinking feeds, however many catfish farmers prefer floating pellets so they can see the catfish feeding response as they come to the surface for food. When compared to the costs of typical livestock feeds, the high nutritional standards seen in aquaculture feeds, as well as the utilization of specialized manufacturing techniques, are reflected in the high costs of most aquaculture feeds. Feed makers' demands to overcompensate quantities of certain critical nutrients, either because absolute requirements are unknown or because uncertain amounts of soluble ingredients may be lost from pellets before they are eaten, drive up these prices.

Oil is removed from the fish before it is dried to make fish meal. Fish is dried using a variety of processes, including hot air or steam. Low temperature (LT) fish meals are becoming more popular in aquaculture feeds as evidence mounts that they are of higher quality in terms of nutritional availability for fish. LT foods are dried at temperatures below 80°C, as opposed to above 100°C for traditional fish meal production. However, the freshness of the raw material is the most important component in determining the quality of fish feed (Birkeland, Rora, Skara, and Bjerkeng, 2004). Oily pelagic fish species including herring, caplin, and anchovy are commonly used to make fish feed. These fish spoil quickly, thus producing high-quality feed for aquaculture requires careful monitoring of the raw materials. This usually entails chilling the fish as soon as possible after catch at sea and minimizing storage time. In the tropics, producing high-quality fish feeds presents unique challenges in terms of competing needs for raw ingredients, spoilage, and subsequent quality control. This has widened the international trade in fish meals, allowing rapidly emerging feed manufacturing sectors like domestic shrimp feed manufacturers in Thailand and Indonesia to import high-quality European and South American fish meals to meet the nutrient specifications of modern shrimp feeds, which are difficult to meet with domestic fish meals. International demand for fish meal could limit the expansion of intensive aquaculture for carnivorous species in the future. Peru, Chile, Denmark, Japan, Iceland, Norway, the United States, and South Africa, which are major producers of high-quality fish meal, are already depleting pelagic fish stocks beyond their ecological limits. Demand may certainly surpass supply before the end of the decade unless new sources can be located that are currently wasted in the fisheries, as shrimp and fish growers compete with the long-established poultry and livestock sectors for a restricted resource. If some fish species, historically used for reduction to fish meal, are utilized in products for direct consumption, improvements in fish harvesting and processing technologies, together with changing consumer demand, may have an impact on future supplies.

2.7. NUTRITIONAL REQUIREMENTS

The nutritional requirements of carnivorous animals limit the amount of plant proteins that may be included in meals for most intensively farmed fish and shrimp. However, some protein sources from agricultural plants, like soybean meal, are employed in diets for both carnivorous and omnivorous animals at variable degrees. Soybean, canola, and other cereal proteins are expected to play a larger role in formulated meals in the future if existing biotechnologies are employed to remove enzyme inhibitors and antinutritional components. Single cell proteins (SCP), made up of bacteria or yeasts cultivated in industrial-scale fermentation systems, may become more popular as fish meal substitutes in aquaculture feeds (Najafpour, 2007). However, most intensive culture feeds already incorporate fish meal as a primary element, and their use in both intensive and semi-intensive aquaculture is quickly rising. In addition to concerns about the future availability of food resources for intensive aquaculture, the current use of aquaculture feeds has prompted environmental concerns. The proximity of several trout farms discharging wastewater into rivers resulted in pollution issues, which were eventually resolved by the installation of industry-wide feed regulations. Wet feeds were banned, and effluent discharge regulations were strengthened. More recently, legislation limiting food conversion values has been introduced. This latest measure puts pressure on farmers to optimize feed management procedures in order to prevent food waste. The experience in Denmark and elsewhere has prompted a thorough investigation into the chemical and biological composition of fish farm effluents, as well as their effects on natural water bodies. The environmental impact of aquaculture is mostly due to feed management difficulties. The most typical cause of difficulties is overfeeding. The physical and chemical nature of waste products, as well as the quantity produced, are affected by feed mix and food conversion values. Uneaten food, excrement, ammonia, and carbon dioxide are the most common waste products connected with feeding. Antibiotic residues may also be present as a result of medicated feed consumption. The water exchange characteristics of the fish or shrimp farm site have a considerable impact on the environmental impact of these materials.

2.8. AQUATIC ENVIRONMENT

In the aquatic environment, feces and unconsumed food provide an abnormal loading of organic material into the farm environment, potentially causing major alterations in water quality. The initial bacterial decomposition of organic compounds added to fish farm water depletes oxygen levels. If organic materials build to substantial levels in the sediment underneath floating cages or on the bottom of ponds, the sediment will become anaerobic and deadly hydrogen sulfide and methane gasses will be released. Suspended solid debris in the water column can injure the delicate gill surfaces of fish and shrimp, impairing respiratory function and create possible sites for bacterial infection. The emission of phosphates and nitrates from fish farms may have long-term consequences. While they are harmless, they can cause water enrichment or eutrophication. This offers favorable conditions for the growth of plants and other organisms. These may compete for oxygen with farmed fish stocks, contributing to the emergence of potentially dangerous plankton blooms in severe circumstances. The leakage of phosphates into the environment is causing great concern. Phosphorus is an essential mineral in fish nutrition, as it aids in development, bone mineralization, and lipid metabolism. Many commercial diets include phosphorus levels that are higher than the known requirements for only a few species. As a result, rather than risk impairing the fish's growth, feed suppliers offer excessive amounts. This is a prevalent issue in the formulation of fish feeds, and it arises from a lack of knowledge about the limited requirements of the cultured species for which feeds are produced. For most farmed species, further research is needed to have a better knowledge of mineral digestion and absorption processes. Many countries have implemented limitations on the scale of fish farms and the phosphorus concentration of farm effluents to reduce the negative effect of fish farm effluent in eutrophication of natural waters receiving fish farm effluent.

Annual catch fishing landings are anticipated to be around 100 million tons (Pauly and Zeller, 2016). This figure is expected to remain steady in the future, therefore any future increase in seafood consumption will have to be satisfied through aquaculture Many factors will influence actual growth over the next decade, including global economics, the availability of vital resources, and the industry's capacity to handle environmental concerns. Large-scale commercial development of intensive culture of species like channel catfish, Atlantic salmon, and giant tiger shrimp, is already extending to marine finfish species like turbot, sea bass, and sea bream, while culture techniques for new aquaculture species like Atlantic halibut, African catfish are in commercial development. On a larger scale, as farmers aim to increase production from semi-intensive cultures of shrimp and prawns, as well as finfish like tilapia, milkfish, and carp, the usage of additional feeds in semi-intensive cultures is growing. Fish have a wide range of feeding behaviors in these various settings. The structural adaptations of their jaws and teeth, the functioning of their digestive systems, and their feeding habits all reflect this. The majority of fish species are carnivorous, meaning they have evolved to eat mostly animal materials. Only a few species are real omnivores, eating both plants and animals, and true herbivores, eating only plants. We may find organisms that utilize practically all forms of organic food sources in nature among aquaculture species, ranging from fish species that prey on other fish to detritus feeders that feed on decaying plant and animal matter. Although we can classify aquaculture animals as herbivores, omnivores, or carnivores in broad terms, strict categories are often unsuitable because some species will graze on whichever supply is most abundant. This is especially true for many species' fry and juvenile stages, where food particle size may be the most important determinant in influencing food intake. Most fish species that are largely herbivorous as adults graze on algae and mixed plankton during their juvenile stages. With few exceptions, fish raised in intensive aquaculture are carnivorous, whereas those raised in extensive and semi-intensive systems are omnivorous and herbivorous species. The eating habits of decapod crustaceans that are commercially important in aquaculture are far less diverse. They have been observed feeding on the flies, depending on the relative abundance of prey organisms. Penaeid shrimps are carnivorous, eating on small animals, but freshwater caridean prawns and crayfish are omnivore, grazing on a variety of plant and animal matter.

The requirement to guarantee that food provided to fish and shrimp is absorbed as thoroughly and quickly as possible, and that the food includes the necessary nutrients to meet the animal's metabolic needs and encourage growth, are both critical considerations in the effective use of feed. Food that is not consumed becomes a potential contaminant, and delaying consumption increases the danger of soluble nutrients leaching. An analysis of feeding procedures reveals that food is being supplied to fish or shrimp in excess of the levels required to match their demand in cases where food wastage arises from overfeeding. In order to determine the best feeding regimens for fish, fish producers must first study hunger and eating behavior. In order to optimize growth and feed conversion values, feed regimens, meal sizes, feeding frequency, and duration of feeding bouts must all be changed. Most fish growers use appetite as the primary criterion for determining feeding requirements. Active swimming and searching characterize the first feeding behavior, which is followed by robust feeding. This is followed by a gradual decrease in the feeding response until the fish stop eating. The fish usually assemble after feeding, swimming beneath the surface. The physiological mechanisms that regulate appetite in fish appear to be comparable to those in higher vertebrates. A short-term response to stomach distension and a longer-term response to levels of particular nutrients circulating in the blood system are thought to operate as regulatory feedback mechanisms. The brain, which is coupled to receptors in the stomach or foregut wall, evaluates the degree of distension as feeding progresses during feeding bouts. Satiation regions in the hypothalamus area (Figure 2.5) of the brain respond to the information received when the stomach is full or half full, and feeding activity is slowed and eventually stopped. The stomach will not generally resume feeding activity until it has passed a considerable amount of its contents into the intestine. Appetitive behavior and the quest for food will be triggered at this time by feeding centers in the hypothalamus, which monitor input from a variety of sources. Changes in the quantities of nutrients-glucose, fatty acids, glycerol, and essential amino acids-absorbed from the digestive tract and circulating in the bloodstream of the fish are also monitored by receptors in the brain. Although several theories have been presented, the physiological mechanisms regulating fish food intake are poorly understood.

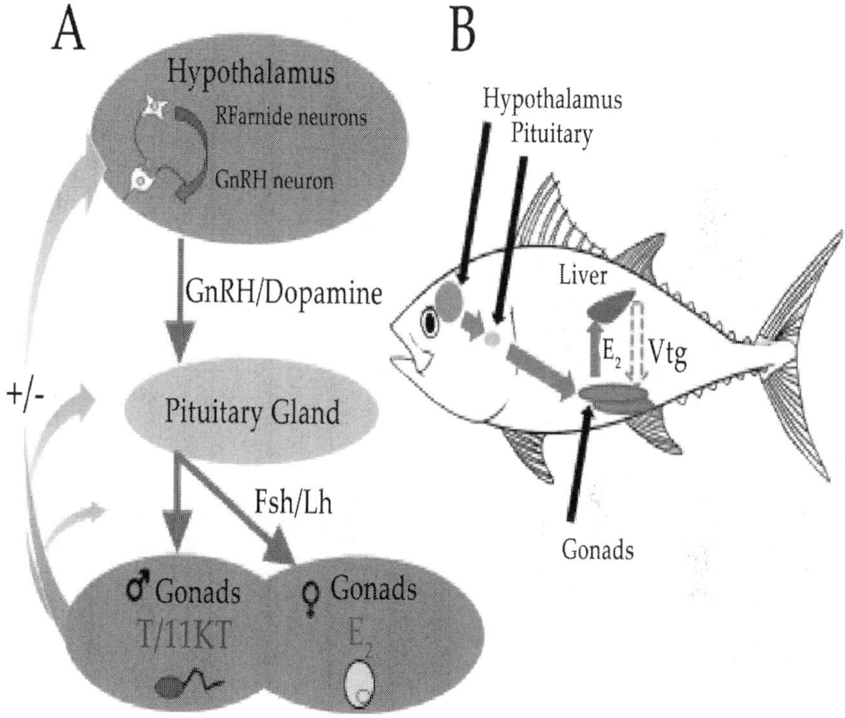

Figure 2.5. Hypothalamus.

Source: https://d3i71xaburhd42.cloudfront.net/4430d40ff56bd7cc079c0f3efc6910ff15bab938/21-Figure1.1-1.png.

2.9. FISH APPETITE

It's still unclear how much stomach distension influences fish appetite, but satiation time and the following return of hunger appear to be the key variables limiting the number of meals a fish will consume each day. It has been discovered that eating behavior is controlled by the metabolic needs of the fish. Fish manage their food intake to suit their energy requirements, according to several studies. Fish adjust for low energy density in their food by eating more, assuming that food is always available. The physical capacity of the gut determines the limitations of such compensation. Fish fed varying energy density diets will grow similarly up to this point, if their diet contains an adequate mix of nutrients. Some species, such as the common carp and some tilapias, have pharyngeal teeth. These compress and decrease food before it enter the colon for chemical digestion. Other species, such as milkfish, have a gizzard-like structure in the foregut that reduces the size of food particles before enzymatic digestion in the long intestine. Carnivorous fish have efficient protein and fat digestion, resulting in high nutritional absorption. However, their digestive systems are incapable of processing or absorbing large amounts of dietary glucose. They don't produce enough enzymes to break down complex carbs like starch, and their digestive tracts are too short and gut clearance times are too fast to enable longer retention times for better digestion and absorption. Starch that has been pre-treated in the feed manufacturing process is more digestible than raw untreated starch, and it is a popular constituent in feeds for most fish and shrimp species, where it serves as a binding agent as well as a source of energy.

The presence of substantial amounts of indigestible elements in omnivorous and herbivorous species' natural diets, such as plant fibers, results in lower levels of digestibility and absorption per unit of food swallowed than in carnivores. However, carnivorous fish can digest and absorb some carbohydrates more efficiently than these fish. Environmental parameters like temperature and salinity, as well as biotic factors like age and size, have little effect on the efficiency of digestion in fish. The many types of penaeid shrimp raised on farms are mostly carnivorous. They eat small organisms that live in or on the surface layer of the pond bottom, such as mollusks, crabs, and polychaetes. They also eat some plant material, especially in the juvenile stages, as well as plant detritus, which contains nutrients generated from the related microbes. In the diets of cultivated shrimp, some foods may predominate. The southern brown shrimp, *Penaeus subtilis*, grew in semi-intensive culture ponds in northern Brazil and showed a preference for benthic polychaetes in terms of diet, eating rhythmicity, and feeding behavior. Polychaetes accounted for 32.5% of overall food intake, while formulated food accounted for 15.6%, according to stomach contents and stable carbon isotope ratios. Sampling of stomach contents over a 24-hour period revealed that applying formulated feeds to the ponds increased intake of natural prey organisms, probably by encouraging foraging activity (Parrish, Deibel, and Thompson, 2009). When additional feeds are utilized in semi-intensive culture, careful monitoring of prey organism populations is essential for feed management. Throughout the rearing cycle, fluctuations in the quantity of prey organisms should be factored into the calculation of formulated feed size.

2.10. RECEPTORS

Sensory receptors on the first chelae detect prey. The food organism is captured by the chelae and delivered to the mouthparts if capture is successful. The maxillae, as well as the first and second pairs of maxillipeds, are designed to hold prey before consumption. The mandibles chop large food items, while non-food things are rejected. Mucus secretions from glands on the paragnaths are mixed with the food mass, and the resulting bolus is forced down the esophagus by the labrum's feeding process. When shrimp are handling and eating natural food items that preserve their shape and integrity, the feeding procedure described above is relatively efficient. This happens when the mandibles crush the food and the mouthparts manipulate it. Pellets that are poorly bonded may be lost by the animal during feeding. While it could be argued that uneaten food helps to the pond's nutritional cycle, this is definitely not the goal of employing costly feeds. Food that is not consumed is wasted, and the consequences for the pond's rearing environment are unexpected and potentially disastrous. Shrimp feeding observations in glass aquaria provide a useful measure of shrimp food quality. Their attractiveness to shrimp, simplicity of manipulation, and stability during handling and consumption can all be studied and compared. Water stability is a crucial element. Shrimp have a harder time manipulating soft, water-soaked pellets. Nutrient loss is quite severe if the pellets disintegrate before or during feeding. Feeding trays allow for observation of feeding activity and pellet stability during the critical first 1–2 hours after feeding, when food should have been consumed.

Food is combined with digestive enzymes secreted by the digestive gland in the stomach before being ground in the gastric mill. In the stomach, food particles are sorted by size. The digestive gland retains larger particles for further processing, whereas small particles pass through. Prior to elimination, indigestible items travel from the stomach into the mid- and hindguts. Proteases, lipases, and sugars are among the digestive enzymes that handle liquid food within the digestive gland. There are also emulsifying agents present. Within the tubular structure of the digestive gland, a variety of specialized cells release enzymes and absorb and store nutrients. Protein and lipid absorption efficiencies are generally high for proteins and lipids, but lower for carbs, depending on the quality of the diet and the balance of nutrients. As the food enters the digestive gland, it is finely divided, which helps to speed up digestion. The foregut's low storage capacity plays an important role in defining feeding regimens. The foregut is estimated to account for 2% to 3% of the total body volume. Because of their limited ability to store food and their quick digestion, actively growing shrimp must feed almost constantly.

Nutrients become accessible to support the metabolic demands of fish or shrimp following the digestion and absorption of food from the digestive tract. When selecting and using various types of aquaculture feeds, it is crucial to have a basic awareness of the nature of critical nutrients employed in them. Much of the research that has gone into current feed formulations have been done on a laboratory scale using juvenile fish for practical reasons. The feed requirements of larger fish and broodstock have received less attention. This endeavor necessitates large-scale, dedicated facilities and must span multiple fish generations. Carbohydrates, lipids, and protein can all be oxidized for energy, while trace minerals and water-soluble vitamins are important coenzyme components in biochemical systems. Essential nutrients are those for which fish and shrimp have a strict dietary requirement and are unable to synthesize from other sources. Animals fed diets deficient in available critical nutrients may experience stunted growth, deformities, or disease. Certain amino acids and fatty acids, as well as most vitamins and minerals, are essential nutrients.

When comparing the nutritional requirements of various cultured fish species, the main variances are in their carbohydrate utilization ability and essential fatty acid requirements. The dietary requirements of diverse cultivated shrimp species have received minimal comparative research. Currently, diets are generally developed using data compiled from the outcomes of research conducted on the group as a whole. Carnivorous fish and shrimp require a lot of protein in their diet. Experiments including feeding high-quality protein to animals in dose-response trials have revealed that carnivorous fish require 40–50% protein in their diets on a dry weight basis, whereas omnivorous fish require 25–35% dietary protein for maximal growth. Younger animals, on the other hand, require more protein to support their faster growth than older animals. Grower diets for older, larger fish typically contain 5–10% more protein than larval, fry, and juvenile diets. As sources of amino acids, proteins are crucial in the diets of fish and shrimp. Following digestion, free amino acids are absorbed from the intestine and transported to the various organs and tissues through the circulatory system.

They can be converted into new proteins for growth and maintenance, or they can be used as a source of energy at the tissue level. The biological value of a consumed protein to an animal is influenced by the profile of its constituent amino acids, as well as their availability following digestion and absorption. While animals can manufacture and interconvert some amino acids, 10 amino acids have been identified as important dietary elements to promote maintenance and growth in fish and shrimp that have been investigated in depth. These are amino acids that animals can't synthesize in large quantities to enable optimal growth. While all fish and shrimp have been found to require the same amino acids in their meals, the amounts required differ by species. Feeding and growth experiments with chemically defined diets were used to determine the amino acid requirements of various fish species.

In order to create and manufacture feeds that include sufficient levels of these critical elements, nutritionists need amino acid profiles of protein ingredients. Fish and shrimp meals often contain high levels of good grade fish meal, which provides an abundance of important amino acids. To meet the requirements of a certain fish or shrimp species, feeds containing high levels of proteins of plant or microbial origin, or protein obtained from animals other than fish, may require supply of critical amino acids. Aquaculture feeds get their necessary fatty acids from dietary lipids. Dietary phospholipids are important structural components of biomembranes and provide a significant source of energy. Other nutrients, including the fat-soluble vitamins A, D, E, and K, as well as natural and manufactured pigments, are transported by dietary lipids. Essential sterols and phospholipids are also found in dietary lipids. Sterols have a variety of biological activities and are used as precursors in the production of vitamins and hormones. Fish lipid nutrition has emerged as a critical research subject in the creation of aquaculture feeds. Both between species and at different life cycle periods of individual species, significant changes in essential fatty acid requirements have been documented. For those species where essential fatty acid requirements have been established, such as salmonids, the focus is on finding the best balance between lipids and other elements in terms of their contribution to the overall energy content of the diet. Consumers are also interested in the impact of dietary lipids on flesh quality and the health implications linked with their roles in human nutrition.

Nature contains over 40 fatty acids. The quantity of carbon atoms in a molecule, as well as the number and position of double bonds between carbon atoms, govern their physical, chemical, and nutritional properties.

A single carboxyl group and a straight, unbranched carbon chain are found in most naturally occurring fatty acids. The carbon chain might have no double bonds, one double bond or monounsaturated fatty acid, or multiple double bonds or polyunsaturated fatty acid or PUFA. Triglycerides are fats and oils that are commonly found in feed stuffs and in the fat deposits of most animals. When one molecule of glycerol mixes with three identical or distinct fatty acids, esters are created. The natural availability of different lipid types present within the food chains of freshwater and marine settings is reflected in the lipid requirements of various farmed fish and shrimp species. The n–3, or linolenic, fatty acids found in the tissues of fish and shrimp are the most abundant. The presence of certain n–6, or linoleic series, fatty acids in the tissues of warm-water freshwater fish is thought to represent intake of lipids from terrestrial sources, where n–6 series fatty acids predominate. Because fish and shrimp are unable to produce n–3 and n–6 fatty acids, they must be given in their diets (Taşbozan and Gökçe, 2017).

2.11. FRESHWATER FISH

Freshwater fish from cold waters have been demonstrated to have an essential demand for n–3 fatty acids, whereas freshwater fish from warm waters require either mixtures of both n–3 and n–6 fatty acids, or fatty acids from the n–6 family alone. Because they lack the ability to extend chains and further desaturate linolenic acid, marine carnivorous animals must be fed highly unsaturated fatty acids (HUFA). Shrimp essential fatty acid needs have received less research than those of fish. They appear to have needs for both lineic and linolenic fatty acids. Crustaceans, such as shrimp, have a nutritional requirement for cholesterol, which they cannot generate from other nutrients, unlike fish. Cholesterol is necessary at dietary levels of 0.5–1.25% of total diet as a precursor of sex and molting hormones and as a constituent of the exoskeleton (Kumar et al., 2018). Oils from marine invertebrates such as squid, shrimp, or clams are high in cholesterol. Phospholipids are also included in shrimp diets. Dietary phospholipids have been found to boost farmed shrimp growth and survival. While shrimp may manufacture phospholipids from other nutrients, the natural rate of biosynthesis appears to be too slow for optimal growth in culture. In extensive aquaculture, the interrelationships between water quality and nutrition are complicated. Water temperatures and dissolved oxygen (DO) levels have an impact on feeding activity, metabolism, and growth, and are thus crucial in determining the types and quantities of feeds employed. Ammonia, which is predominantly expelled through the gills of fish and shrimp, may have direct harmful effects on these animals, whereas phosphorus, nitrogen, and organic solids may have an indirect influence on fish and shrimp through long-term environmental degradation. Some hazardous substances are excreted as a by-product of regular metabolic processes, while others are discharged as a result of improper feeding practices or poor food quality.

In actively feeding fish, nitrogenous wastes, mostly dissolved ammonia, and feces containing indigestible components are produced almost continually. Ammonia is very hazardous to fish and shrimp in its unionized state, whereas feces thrown into the water quickly break down, contributing to the water system's suspended and settleable solids deposition. The amounts of ammonia and feces are normally managed in holding systems with suitable water exchange rates by continuous dilution by inflowing water. In static pond systems, waste products enter the natural pond ecosystem's nitrogen, carbon, and phosphorus cycles. These chemicals are degraded by microbial activity and their elements are released into the water under ideal conditions. They become available as nutrients for algae here, and in well-managed pond systems, nutrient cycling contributes significantly to the pond's primary productivity. The discharge of carbon, nitrogen, and phosphorus-containing farm effluents into receiving waters and coastal basins may cause detrimental eutrophication or enrichment. These changes may lead to conditions that encourage the growth of algal blooms, which can have severe consequences for fish and shrimp supplies. While some water quality criteria can be controlled during ordinary farm operations, the effective limits for control are established during the initial selection of the farm site as well as the planning of the farm layout and water supply. The consequences of feeding fish in intensive systems on water quality have garnered a lot of attention in recent years, especially in northern Europe and North America, where intensive aquaculture sectors have developed within strict legal frameworks for water pollution prevention. Initially, difficulties were mostly investigated in terms of efficient chemistry, but more recently, a larger emphasis has been placed on addressing problems at their source by investigating nutritional solutions. Both the nature of feed ingredients and the overall nutrient balance of feeds have undergone considerable modifications. These adjustments, together with advances in feeding procedures, have resulted in significant reductions in pollution loadings in some industries and pave the way for others.

Each fish and shrimp species has a recommended water temperature range for optimal eating, metabolism, and growth. Obligate poikilotherms, such as fish and shrimp, have little or no ability to control their body temperatures independently of their surroundings. When the temperature of the water changes, the fish's body temperature quickly adjusts to the new temperature. Because heat is transferred through the body wall and gills, temperature changes occur more quickly in tiny animals than in bigger organisms. While fish's body temperatures are very similar to those of their surroundings, they are extremely sensitive to temperature fluctuations, with some being able to sense temperature changes as little as 0.5°C (Forster, Hirst, and Atkinson, 2012). Fish can seek their preferred temperature zones in natural waters, but the temperature ranges available to fish in aquaculture systems are generally limited. However, no single species can survive across the full range, and each species has its own temperature range with upper and lower limits. Coldwater and tropical species have the highest temperature tolerance in aquaculture. Salmonids and some marine fish species are cultivated in northern climates, where part of the production cycle may take place under ice or at temperatures approaching zero. Fish and shrimp raised along the equator, on the other hand, spend the most of their lives growing at temperatures above 30°C (Rubio et al., 2019). The optimal temperature range for any farmed species is particularly important in aquaculture. The optimum temperature range is defined as the temperature range in which feeding occurs without inappropriate behavior due to thermal stress. A narrow growth optimum may also be possible to define. Fish suffer from heat stress outside of the ideal range and die at the upper and lower lethal limits. There is a similar increase in metabolic rate as water temperature rises within the fishes' optimum range. Higher metabolic rate leads to faster food transit through the fish's digestive system, which leads to increased hunger and food intake. Assuming that food is abundant and that DO levels are not a constraint, energy intake achieves maximum and growth rates reach their peak.

Because this is a light-dependent mechanism, oxygen is created during the day. Photosynthesis stops at night, but plants continue to breathe, receiving oxygen and emitting carbon dioxide. This whole process can result in oxygen and pH cycles. Depending on plant populations, conditions may shift between sub-saturation and supersaturation. Plant activity upstream of the farm will influence dissolved oxygen (DO) levels in farm ponds in fish farms fed with flowing water, rivers, or streams, and similar effects will occur in fish ponds or cages where considerable populations of algae are present. DO fluctuates between 2 mg/L and 3 mg/L throughout the day. In productive fish ponds with little or no water exchange, however, fluctuations can reach 7–8 mg/L (Ozbay, Blank, and Thunjai, 2014). Fish are extremely stressed at these levels, and their growth and food conversion are hampered. Microorganisms that decompose organic matter (OM) in the benthos at the bottom of fish ponds or below fish cages may likewise place a high demand on DO. The biochemical oxygen demand (BOD) is a commonly used metric for determining the impacts of organic loadings in aquatic systems. The aerobic breakdown of surplus feeds and other organic elements in the sediment accounts for a large portion of the BOD in intensive shrimp ponds. The sediments in intensive shrimp ponds absorb up to three times as much oxygen as water column organisms and shrimp biomass combined. Knowledge of DO levels and potential changes at the farm site is critical for optimizing feeding regimens for any cultivated species. Because oxygen requirements are high and oxygen has a limited solubility in water, oxygen levels may decrease near to or below the minimal threshold. As feed volumes increase, intensive shrimp ponds transition from autotrophy to heterotrophy, with net positive DO values. Fish and shrimp initially display signs of stress when there is a lack of DO.

Fish may swallow air at the surface in extreme conditions of oxygen deficiency. Shrimp move into shallow water or swim near to the surface when there is a lack of DO. Shrimp farmers use a flashlight to examine the edges of shrimp ponds at night. Shrimp assembling on the shallow edges is seen as a sign of low DO levels. Keeping DO levels above the needed minimum did not boost food intake, food conversion efficiency, or growth rates in rainbow trout or channel catfish tests. Food conversion values were revealed to be negatively affected by changes in DO levels over the minimal threshold. Energy is required for fish and shrimp to move, find, and digest food, grow, and maintain body functions. Energy contained in their feed must be transformed into usable energy through metabolic processes. All oxidative metabolic activities require oxygen as well as an organic substrate. These oxidative processes are more biochemically effective than anaerobic mechanisms in animals with a relatively consistent oxygen supply. A lot of factors influence oxygen consumption rates, including body weight, water temperature, and amount of exertion. Larger fish consume more oxygen per hour than smaller fish, whereas smaller fish consume more oxygen per unit of body weight. Fish require more oxygen at higher temperatures and activity levels than at lower temperatures and activity levels. Fish species have demonstrated significant adaptation in their capacity to exist and function in waters with varying levels of DO. Warm-water species, on the other hand, can live in waters with lower DO levels than cold-water ones. Salmonids, for example, have high DO needs, with a tolerance threshold of 5–6 mg/L (Remen et al., 2013). Tilapias, on the other hand, have lower total requirements because they live in slow-moving tropical rivers, lakes, and ponds. A DO content of 3 mg/L is required for tilapia. Fish exchange breathing gasses over the gill lamellae's surface. Carbon dioxide is expelled while oxygen is absorbed. Gases are transferred by diffusion across the thin lamellae membranes that divide the blood circulatory system from the water flow via the gills. The partial pressures of oxygen and carbon dioxide in the water, pH, temperature, and the activity level of the fish all affect the oxygen concentration of the blood.

The rate of oxygen intake increases after eating. The rate of oxygen consumption climbs to a peak shortly after eating, then gradually decreases back to the pre-feeding resting level. The specific dynamic action (SDA) refers to the metabolic activities connected with the assimilation and absorption of nutrients and refers to an increase in metabolic rate following the ingestion of food. Although the effects and a variety of factors that influence the SDA in fish have been described, the specific causes are unknown. Feeding fish to satiation can result in such high oxygen demand, essential to fulfill SDA, in extreme circumstances at high temperatures, when DO levels in water are decreased and resting metabolic oxygen consumption by fish is high. The maximal rates of oxygen consumption produced by eating are around double those of resting fish under normal conditions. The time it takes for SDA to have an effect on oxygen consumption depends on a number of parameters, including water temperature, meal size, food composition, and fish weight. Peak oxygen demand occurs within 12 hours of eating a meal, and some research suggests that the length of the effect is related to the rate at which food passes through the gut. SDA may also be associated with appetite and food intake regulation (Volkoff and Rønnestad, n.d.).

2.12. ANALYSIS

Electronic meters and probes, when combined with data loggers, can provide farmers with continuous recordings of O and other water quality parameters. The majority of issues connected with DO in temperate regions occur during the warmer summer months, and are caused by a combination of causes. The oxygen retention capacity of water is lowered at higher temperatures, which usually coincides with periods of active growth, high stocking densities, and increased feeding levels. During the summer, the rapid growth of microbes, algae, and higher plants adds to the oxygen demand. Critically low DO levels can be caused by a combination of any or all of these reasons. These elements are present all year in the tropics and subtropics. When diurnal DO changes are generated by plants' photosynthetic-respiratory activities, DO levels should be measured before day-break and monitored until safe levels are established before feeding, keeping in mind that fish's oxygen requirements peak several hours after eating. Under these conditions, food consumption should be closely checked and feeding rates changed as needed to avoid wastage and any additional contribution to oxygen demands. Culture ponds experience oxygen depletion more frequently than tanks, raceways, or cages. DO levels fluctuate due to low water exchange rates, plant photosynthesis, and the respiratory activity of the pond biomass. When high stocking densities are supported by high rates of water exchange in tank and raceway systems, oxygen depletion is most usually caused by a system failure in the water supply or aeration system, or contamination of the water supply. Currents, wind, and fish movements all contribute to water exchange in cages or pens. Fluctuations in DO levels can occur as a result of plant photosynthetic activity or, in extreme conditions, phytoplankton blooms.

While producers of fish and shrimp feeds have made significant progress in developing extremely appetizing and digestible diets, waste products are an unavoidable component of the feeding process and must be accounted for in farm management processes. Poor feeding techniques and erroneous estimates of the biomass of animals being fed are frequently responsible for high levels of wasted uneaten food. The amount of uneaten food varies depending on the type of food used, which might be wet, moist, or dry. Food wastage estimates for cage sites are higher than for ponds, tanks, or raceways. Detailed studies in Scotland estimated that uneaten food from rainbow trout and salmon farms accounts for up to 20% of the food fed (Wang, Olsen, Reitan, and Olsen, 2012). This is a large number that has a substantial impact on the profitability of cage-rearing fish. Waste detection devices are being developed in response to the economic and environmental concerns involved with food wasting at cage sites, in an attempt to minimize or mitigate the problems related with overfeeding. The nature of the shrimp pond ecosystem and the availability of natural food organisms make accurate measurement of uneaten food difficult. In the Philippines, an estimate of uneaten food for gigantic tiger shrimp farms was provided as 15% (Dauda et al., 2019). A comprehensive investigation of karuma shrimp ponds in France revealed much greater amounts of food waste, which was linked to population densities of preferred natural food organisms. The fate of uneaten food is determined by the holding method used. The majority of uneaten food remains in stagnant ponds with poor water exchange rates. Feeds used in fish and shrimp cage rearing are typically high density and drop quickly below the cage if not consumed. Food pellets break up in the silt and are consumed by other species such as polychaete worms, mollusks, and insects. Bacteria decompose waste food that has not been consumed. This will increase the oxygen requirement under aerobic conditions. When oxygen is scarce, the sediment becomes anaerobic, and hydrogen sulfide, produced by heterotrophic bacteria, can build up to dangerous amounts. Hydrogen sulfide is a highly soluble gas that is harmful to fish and shrimp at extremely low concentrations, and it can cause serious difficulties in fish cages and culture ponds when there is insufficient water exchange to dilute it to safe levels.

Feces can include up to 30% of the dietary carbon ingested and 10% of the nitrogen consumed. As a result, enormous amounts of waste organic elements are released into the farm environment. Feces can either settle to the sediment or break apart in the water column, increasing suspended solids levels. The suspended solids concentration of water is also influenced by dust particles from poorly bonded feeds. The fragile gills of fish and shrimp can be damaged by high quantities of suspended particles. They may cause physical harm, increasing the risk of bacterial infection, or they may physically obstruct water movement across the gill surface, limiting respiratory function. With extensive feeding and high stocking densities, this type of pollutant loading is especially significant in intensive fish farming. As a result, stocking and feeding practices should be tailored to the species' tolerance for suspended solids. Fish have a wide tolerance range; readings of 25 ppm are generally considered safe. Depending on the composition of the solids, gill structures may be protected at levels between 25 and 88 ppm (Goddard, 1996). When suspended solids levels exceed 80 ppm, most species are vulnerable. The amount of solid waste created during fish culture has been examined the most in the freshwater production of rainbow trout. For each ton of fish produced annually, estimates from cage, pond, and tank rearing systems range from 0.52 to 0.65 tons of solid waste. The results from different farms differed significantly, owing to the various feeding levels and management strategies used on the farms evaluated. Estimating the volumes of uneaten food, excrement, and soluble wastes has also been done using mass balance calculations. The total loadings of waste nutrients are calculated using this method by subtracting the amounts of nutrients provided from the amounts retained in fish or shrimp carcasses. According to tilapia nitrogen balance estimates, 98 kg of nitrogen is lost to the environment for every ton of tilapia produced. This estimate is based on a 3% nitrogen content for tilapia and an 8% nitrogen content for a specific tilapia diet.

When stocking densities and feeding levels are at their peak, harmful elements in organic sediments deposited in semi-intensive or intensive fish and shrimp ponds generally approach critical levels. Fertilizers, plankton, and sediments carried in the water supply, in addition to uneaten food and feces, all contribute to silt buildup. However, excessive feed consumption is the most prevalent cause of pond water quality degradation and the accumulation of hazardous materials. Treatments for pond bottoms can only be done efficiently between production cycles, and usually entail a combination of flushing, drying, and removing excess or anaerobic sediments. Before the sediments settle, ponds must be cleaned soon after harvesting. To dislodge and flush sediments, rakes, drag chains, and hydraulic jets are frequently utilized. In intensive culture ponds, the top 5–20 cm, reduced layer of silt may need to be removed or replaced, either manually with dredges or mechanically with earthmoving equipment. Excretory and waste products can have consequences on waters receiving farm effluent, in addition to their effects on animals in culture systems. This is a key issue in feed management since effluent quality is directly linked to feeds and feeding practices and is controlled in many countries under water pollution control regulations. The regulations that control aquaculture effluents are reviewed on a regular basis as new information regarding the physical and chemical natures of fish farm effluents and their effects on receiving waters becomes available. Aquaculture is controlled in most countries within the scope of existing legislation designed to control industrial and agricultural effluents. Phosphorus (P), nitrogen (N), OM with its accompanying BOD, and suspended particles are the principal contaminants linked with fish farming. Limits on the discharge of contaminants from fish farms are often set in nations with effective environmental controls.

CHAPTER 3

FINANCIAL MANAGEMENT AND ECONOMICS OF FISH FARMING

CONTENTS

3.1. Introduction 56

3.2. Business 57

3.3. Marketing Problems 59

3.4. Evaluation 60

3.5. Consumer Demand 62

3.6. Restrictions 65

3.7. Credit Reserves 71

3.1. INTRODUCTION

The bulk of aquaculture enterprises (Figure 3.1) demand significant operating and investment resources. Obtaining sufficient finance is generally one of the most difficult challenges in beginning an aquaculture business. Farms and processing plants with insufficient capital rarely survive. The amount of capital required to operate at a high degree of efficiency, as well as the sources of that money, require careful consideration and planning. The initial cash required to purchase land, construct production facilities, and purchase equipment is the first step. New roads may need to be built, electric power lines may need to be laid, or more infrastructure may be necessary, all of which will increase the overall amount of investment capital required. The cost of marketing facilities must be factored into the budget. A water supply system to fill carrying trucks will be necessary. Depending on the type of items sold, an ice machine may be required. Trucks, tanks, oxygen systems, and loading equipment will be required if the farm will perform its own hauling and transporting. Overall, the investment capital for an aquaculture business will normally be several 1,000 dollars per acre of output for pond systems, and can range from \$0.30 to \$7.00 per pound depending on the species and production technique (Bostock et al., 2010). Yearly fixed costs are substantial in an aquaculture business due to the large level of investment capital required. Producing fish at an intensive level with high yields is the greatest strategy to reduce the fixed cost element of the cost of production since high yields spread the annual fixed costs over a larger volume of production and lower the cost per pound of production.

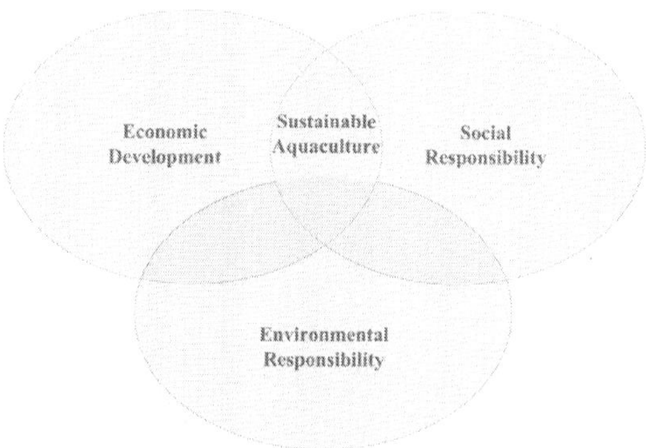

Figure 3.1. Aquaculture enterprises.

Source: https://www.fishfarming.com/sites/default/files/images/services/sustainable%20aquaculture.jpg.

3.2. BUSINESS

For aquaculture firms, operating capital requirements are frequently as high as initial capital requirements. This is frequently because high yields are required to reduce annual fixed expenses per pound and maintain production costs competitive. To achieve big yields, you'll need a lot of fingerlings (Figure 3.2), a lot of feed, a lot of energy for aeration, and a lot of other facilities like labor, repairs, and maintenance, and gasoline. Operating costs for pond systems range from $2,000 to $5,000 per acre, whereas intensive systems such as raceways and indoor systems can cost anywhere from $33,000 to $150,000 per acre.

Figure 3.2. Fingerlings.

Source: https://www.mainstreamaquaculture.com/wp-content/uploads/2019/08/2018-1-1024x512.jpg.

The fact that some varieties of fish raised do not reach market size in one growing season adds to the operating costs. As a result, a prospective fish farmer must often plan on operating the business for more than a year without getting any revenue. Careful financial planning and effective communication with lenders are essential for obtaining sufficient funds to expand the business until it reaches full production capacity. The large capital requirements of many aquaculture firms result in significant financial risk. The potential for profit is frequently associated with a number of hazards, and huge quantities of money invested in an aquaculture business can quickly be lost. Appropriate and thorough planning, monitoring, and assessment of the aquaculture business's economics and finances throughout its life is the greatest way to avoid such losses.

Aquaculture firms demand a high degree of management due to the large amounts of investment and operating capital required, as well as the intense nature of aquatic animal production. When aquatic animals are kept in a production unit, the quality of the fish's growing environment must be constantly monitored. It demands careful and continual monitoring to maintain acceptable levels of oxygen and other essential water quality indicators, as well as to avoid difficulties connected with the breakdown of waste products in the system. When animals are kept in close quarters, diseases spread quickly. As a result, care must be taken to monitor the animals' health and take appropriate action when disease symptoms appear. Direct-sale farms will necessitate a high level of management attention to marketing responsibilities and operations. Even farms that sell straight to a processing plant must have management who are familiar with processor needs. To minimize dockages from fish that do not meet criteria, these measures include quality standards and delivery requirements, among others. Top managers guarantee that their farm is regarded as a preferred supplier, delivering high-quality fish on schedule and within specifications. Throughout the manufacturing process, management must keep a constant eye on costs and production efficiencies. This includes tracking labor efficiency, feed usage, electricity and fuel usage, and equipment use and care. Feed, for example, is typically the most expensive component of aquaculture crop production. Careful and suitable feeding will result in a higher feed conversion ratio and more pounds of fish produced per pound of feed used. Similarly, rather than turning on more aerators than are needed at one time, intelligent use of aeration can supply enough oxygen levels by turning aerators on progressively as needed.

3.3. MARKETING PROBLEMS

The marketing problems of beginning a new aquaculture business are frequently bigger than and ultimately more essential than the production challenges. Individuals who manage and operate successful aquaculture firms spend as much time researching marketing alternatives and trends as they do focus on production efficiencies. Prospective aquaculture business owners should consider a number of broad trends and difficulties. The majority of aquacultured species were once primarily wild-caught species. Many of these species have established markets and demand that were dependent on supplies from catch fisheries in the past. Regional preferences for wild-caught versus farmed fish differ. It is critical to comprehend these preferences in the market for the company's product. When farmed products become accessible on the market, it frequently has to compete with already well-established wild-caught products. However, the cost of producing farmed fish, particularly in the early stages of a business, necessitates a higher price than wild-caught fish (Bertheussen, Dreyer, and Reiertsen, 2020). In order to acquire a price that covers production costs, a latest product must typically be differentiated from wild caught and other similar items on the market. In the previous few decades, the seafood business has changed dramatically. In order to plan for successful marketing strategies, present and emerging trends must be carefully evaluated.

While dynamic markets such as those for seafood can be difficult, they can also provide opportunity for entrepreneurs. Shrimp and salmon businesses around the world, for example, have benefited from increased global commerce and technology. These aquaculture segments have dominated the global shrimp and salmon markets. In recent decades, food marketing has undergone significant changes, resulting in changes throughout the supply chain (Valenti et al., 2021). The advent of substantial market strength at the level of the huge hypermarket discount retail sector, epitomized by corporations like Wal-Mart, has been the driving force. Wholesalers and food service distributors have become more concentrated as a result of this concentration of market power. This has raised the demand on growers to either consolidate by vertically integrating to gain market dominance, or form cooperatives or other types of groups to compete. Start-up aquaculture enterprises must first determine which market they intend to target. The entire marketing plan must also identify the competition and the distinctive place in the market that the company's product will occupy. The product must be adequately defined and aligned with its market positioning for the intended customers. The size of the market, long-term price trends, and distribution patterns of related products should all be carefully considered. The marketing strategy must include a promotion and advertising strategy. Even the tiniest aquaculture farms must have a strategy for publicizing their products. Promotion is a method of disseminating information about a product's features, pricing, and why a customer should buy it. Market channels that are appropriate and successful must be built. The amount of time required to transport fish to markets must be determined, and the business plan must include appropriate employees. The maximum number of round trips that can be made safely is an essential consideration. If the farmer does not want to transport his or her own fish, he or she will need to establish partnerships and agreements with a wholesaler or distributor.

In the United States, most farms are owned by a single person and are classed as sole proprietorships. The farmer is self-employed and holds legal ownership of the land in a sole proprietorship. This is the most straightforward corporate structure, but it also carries the most risk. In a single proprietorship, risk arises from the owner's exclusive responsibility for any debt obligations or accidents. Furthermore, the farmer's obligation is not limited to the amount invested in the enterprise. Due to severe bad situations, the farmer may lose his or her land and home. To acquire access to extra resources such as land, equipment, labor, or management, some farmers form partnerships with family members or others. Partnerships can be general or specific. In a general partnership, all partners share ownership, control, and responsibility. Limited partners participate in the company's profits and losses but not in its management. The limited partner contributes resources like capital, while the major owner retains authority over management choices. Some aquaculture segments have vertically integrated and developed into corporations. Shares of stock provide capital, and stockholders, the board of directors, and the officers give management and control in a business. The board of directors establishes policy, while the officers oversee the company's daily operations. While stockholders own the company, they are not individually liable for the company's activities.

3.4. EVALUATION

Land, existing ponds, wells or other types of water supplies, farm equipment, and structures should all be included as part of the assessment. The evaluation must be comprehensive and detailed. For example, the individual may have sufficient land, but present or upcoming zoning laws may prevent it from being used for aquaculture. Analyzes of water and soil should be performed to determine the species' compatibility. Some freshwaters have enough salinity to support crops that can tolerate low salinity, such as marine shrimp. Backup equipment, such as generators for power outages and backup aeration equipment, must also be available in the event of a breakdown. The availability of sufficient quantities of seedstock is also a resource. The availability of labor resources can determine whether a new business succeeds or fails. Any family labor that is available to assist with the farm should be considered in the labor availability assessment. A realistic assessment of the local labor supply and the availability of appropriately skilled employees for the aquaculture firm must be made. Aquaculture firms frequently require more skill than other types of agriculture, and workers' capacity to handle the extra tasks must be carefully assessed. Workers who cannot swim or are frightened of water, for example, may find it difficult to acclimatize to working in close proximity to it. During the main growing season, much aquaculture necessitates lengthy and unpredictable hours. Workers may or may not be interested in working such long hours. The amount of machinery on the farm necessitates a lot of upkeep. An aquaculture business will either need to hire a mechanic on the farm or be prepared to pay greater repair and maintenance charges. It is necessary to examine the availability of managerial resources. The owner's level of competence and skill in managing the aquaculture business's production, marketing, and financing must be assessed honestly. If the owner excels at aquaculture but struggles with financial analysis, an accountant or financial analyst should be hired, contracted, or retained. Similarly, if the owner has good business abilities in marketing and finance but no experience with aquaculture, appointing a production manager with adequate aquaculture skill and expertise is critical.

The operational line of credit must be established to allow the business to continue running during the period when it first starts to earn profits. Depending on the type of business, it could take 2–3 years for sufficient money to be created from the aquaculture firm (Yue and Shen, 2022). A significant amount of investment capital must be accessible to provide risk-mitigation facilities. This includes having enough equipment redundancy to overcome power outages, malfunctions, and unanticipated extended periods of bad weather. Maintenance costs must be factored into budgetary planning. This includes sufficient funds to replace all equipment and facilities as needed. This, in turn, will be determined by the individual's financial situation, collateral availability, and general credit worthiness. The species chosen and their product forms are important considerations.

These must correspond to the anticipated price range and quantity demand for that product type and species. When choosing which species to raise, consider whether there will be competition from imported species, capture fisheries, or both. Diversifying farm production with multiple species also helps to spread the risk of market price drops for a single species. Matching the species to be produced, the production technology to be employed, and the volume and scope of operation to the available markets and resources is the key to beginning a successful aquaculture business. Misalignments are more likely to lead to corporate failure. If the owner is unlikely to be able to obtain sufficient funds to build and operate a large farm, it is preferable to rethink the business plan and establish one that can be implemented with the monetary resources available. Under-capitalization is a common cause of aquaculture farm and processing facility failure. Financial failure occurs when expected and real capital requirements differ.

3.5. CONSUMER DEMAND

Consumer demand variables that are quantitatively examined can provide insight into how some of these components interact. Economists calculate demand by regressing the price of a product against independent variables such as the number of units sold, incomes, and tastes and preferences, among other things. Economists then calculate elasticity, or the proportional change in quantity demanded in response to a change in price, using the estimated demand relationships. Elasticity is essential in market discussions because if demand is elastic, a price rise will result in a relatively bigger fall in the amount desired, lowering total revenue for the producer. When demand is inelastic, however, a price increase may still result in a lesser amount demanded, but the quantity response is proportionately smaller than the price change. When this happens, the grower's total revenue will rise as prices rise. In order to successfully command the higher prices that are usual in niche markets, inelastic demand is typically required. One of the characteristics of inelastic demand products is that they have few close substitutes. One of the most important aspects of producing a product for a niche market is to set it apart from the competition.

Efforts to develop a differentiated product must take into account where the product should be placed in the marketplace. This selection must take into account production costs, competitiveness, consumer impressions of the product, and products that are closely competing. This decision can be aided by a product-space map. Products can be positioned as either a high-quality, high-priced product or a low-quality, low-priced product. The idea is to match the price to the consumer's expectations. Certain market segments are willing and able to pay high prices for a high-quality product. The species, product form, packaging, and pricing all contribute to product positioning. Any element that adds value, such as a spice packet, marinade, or packaging size, can influence how a product is positioned and differentiated from competitors. Aquaculture farmers, especially those on a smaller scale, typically prefer to target specialist markets because prices are often higher. Specialty marketing is the decision to make a high-quality product in order to command a high price. Equally essential is how the individual product exceeds the competition in terms of representing those traits. Over time, successful items attract competitors. Other companies will learn to make similar products, and the product concept and line may be expanded. When this happens, the product loses its uniqueness. The company must be ready for the competition that will arise if the new items are successful. Every product that succeeds in the marketplace goes through a life cycle. As clients become aware of the product and begin to experiment with it, sales climb rapidly in the early phases following its launch. However, the product eventually matures to the point that it invites competitors. As the volume of sales approaches a maximum, the rate of growth of product sales begins to slow. After that, product sales start to fall. Companies with mature products must either start looking for new markets that can be either new geographic markets or new market categories or start developing new products. The product's price range in the market, as determined by the company, must be sufficient to pay not only production but also marketing costs.

Large-scale growers must find a high-volume market outlet, such as a processing plant. Before building production facilities, it is critical that a potential grower take time to speak with processing plant purchasers. When considering sales to a processing plant, there are a lot of crucial factors to consider. Some plants offer harvesting and transportation services, while others rely on the farmer to make the necessary arrangements. Contracting arrangements between processors and fish farms vary greatly. The farmer should inquire about delivery volume needs as well as seasonality considerations that may affect delivery. Size considerations are significant. Some plants, for example, may not pay for fish that are smaller or larger than a certain size range, while others may deduct a percentage from the overall load if the proportion of out-of-size fish exceeds the plant's requirement. The grower must fully comprehend the processor's quality requirements, flavor, and meat quality expectations. Payment frequency to growers and the average time between delivery of fish and payment receipt should be compared across plants. Some jurisdictions require processing companies to post bonds to ensure that farmers are compensated for the fish they deliver. The payment history of the processor should be examined.

To be successful, small-scale producers will need to find higher-priced alternative marketing outlets. This is because smaller farms have greater production costs than larger farms due to economies of scale. Processors' payments to farmers frequently fall below the break-even point for small-scale farms. Live sales with custom processing, fee-fishing or pay-lake operations, sales to local grocery stores and restaurants, and sales to live haulers are examples of alternative market channels for small-scale farmers. Live fish sales can be a way to get better pricing in locations where people eat fish often. The capacity to process fish according to customer preferences may appeal to a wider audience. Before creating this type of marketing campaign, take into account state and municipal health rules, permits, and Hazard Analysis of Critical Control Points or HACCP strategies (Hoque and Myrland, 2022). Fee-fishing or pay-lake enterprises essentially sell their consumers a recreational opportunity. Fee-fishing may provide a viable market outlet for farm-raised fish if it is located within 30–50 miles of a major population area. Fish for stocking are also purchased by pay lake operators. Increased interest in state game and fish agencies' urban and community fishing programs may give marketing opportunities for fish growers. Fish shipping arrangements with a live hauler may be required for sales to pay for lakes. Unless restaurant workers clean the fish, sales to local grocery stores and restaurants need on-site processing. Only the owners of high-end seafood restaurants will typically buy whole fish to be cleaned by their staff. Before establishing this type of marketing outlet, take into account state and municipal health standards, permits, and HACCP protocols. Targeting any of these sales locations necessitates precise calculations of volume, size preferences, and sales costs. Furthermore, state and local rules must be thoroughly analyzed, as well as the implications for business expenditures.

All data gathered during the marketing plan development process should be meticulously structured and filed. The state of marketing evolves with time. Year after year, the marketing plan and the data that underpins its assumptions and targets will need to be revised. Maintaining meticulous records that serve as the foundation for each choice will make it easier to adapt the plan as situations change. The long-term series of product pricing over time in various markets will be one of the major pieces of data to be kept. It will be crucial to compare data from groups that show growth and those that show losses in numbers and income over time. Key patterns can be forecasted using this data throughout time, allowing for timely modifications in the company's overall marketing plans. The majority of modern aquaculture operations are capital-intensive. A documented business plan, updated annually, provides a framework for continuous company examination and evaluation. It must thoroughly combine the marketing, production, and financial aspects of the business. The strategy must be consistent throughout all of its component analyzes. Furthermore, the plan must include both short-term and long-term objectives. It is easier to identify the cause of problems and potential remedies with a written plan in place, making it easier to be prepared for times when things go wrong. Appropriate business planning can assist avoid mistakes and reduce risks involved with funding, producing, and promoting the company's products. It takes months to create a decent company plan. It is a well-organized and structured document that assesses the market potential for the products produced, investigates the manufacturing technology chosen, and assesses the company's financial performance.

It's common to mix up business plans with business loan requests. Regardless of whether the capital in the business is borrowed or given by the owner, all businesses should have a solid business plan. Many potential and existing fish farmers report having difficulty acquiring finance for their aquaculture enterprises. Lenders are frequently unfamiliar with aquaculture operations and the market possibilities for aquaculture products. Lenders are concerned about the unpredictability and risk connected with aquaculture enterprises in other cases. These issues usually result in loan applications being denied. Even if a lender understands aquaculture and its potential, a poorly designed business loan proposal will be rejected. A high degree of documentation is required for a successful loan application for an aquaculture operation. A business plan's end result is an in-depth understanding of whether or not the planned business will succeed. It will show you where the biggest issues are and which solutions are most likely to help you overcome them.

3.6. RESTRICTIONS

New restrictions that affect aquaculture enterprises can be an external risk. Regulations can raise costs, limit market access, or have other negative consequences for businesses. Each year, with an eye on the external risk most likely to damage the business, careful planning allows for changes and adjustments to minimize negative consequences. Paying attention to the changing environment might sometimes lead to new chances. The company's internal strengths and shortcomings must also be assessed. A company's strengths in skillful management may be offset by a shortcoming in its ability to borrow further cash. Each year, these strengths and weaknesses must be evaluated in terms of their impact on the overall plan. Internal strengths and shortcomings should both be evaluated. A small business with a greater cost of manufacturing, for example, may be better positioned to establish higher-valued niche markets. A farmer with experience spawning a difficult-to-spawn species of fish may have an edge as a hatchery supplier, whereas another company with enormous amounts of land may focus on grow-out. A business's vulnerability can include aging employees who are unable to offer the physical labor required, or having too many people to function profitably. A company's strength could be thorough market knowledge, exceptional engineering and maintenance skills, or financial analytical expertise. Internal analysis must take into account available financial resources for market research as well as any new investment or operating capital requirements.

Every company may have a marketing strategy and plan that tackles product, price, promotion, and location. These are factors that the company can influence. The farmer, for example, chooses which species and sizes of fish to cultivate as a product to sell. The geographic market, or where the farmer will sell the fish, is referred to as "place." Choosing whether to sell on the farm, haul to a processing plant, or sell to other farms is the "location" decision. The farmer selects how to promote the product through advertising. The farmer also determines the price at which his or her goods will be sold. Market conditions will determine whether or not the product is sold. It's also crucial to know when to sell the product. Because the main crop of bait fish is sold in the spring rather than the fall, this farm will face serious financial difficulties. Because the company did not have an acceptable marketing strategy from the start, it will fail. The marketing plan and strategy must specify which market groups the company will target. Diversifying production, for instance, will almost certainly increase costs. As a result, extensive study is necessary to identify the most profitable market sectors, those with the highest overall potential for accomplishing the company's goals. The corporation must then focus its efforts on developing, producing, stocking, and marketing products that are most likely to succeed in that market sector. The selection of products and product lines must be developed concurrently with the selection of target markets in the company's market plan. A segment must be of sufficient size, with potential for further growth, not overcrowded by competition, and have an identified need that the company can satisfy uniquely. A product with a high production cost must be of adequate quality to command a high enough price to be viable.

Clearly, the target market for such a product would be one in which buyers not only value the product's specific features, but also have sufficient income levels to pay the requisite price. The market segment chosen must have enough users to generate the number of sales required to cover the costs of product development. A successful business and market strategy requires the identification and selection of items and product lines for the company. Several catfish processing companies, for example, provide a marinated filet product line that includes lemon pepper, Cajun, and other flavors. The nugget, steak, and whole-dressed product lines are unique from the marinated filet line. Companies with a single product line may have cheaper production costs due to efficiency, but they may also be more vulnerable to market risk. A corporation can spread the risk of changing market and economic situations by having differentiated product lines. Some companies' business plans may allow them to expand and diversify their product lines, while others may not be able to afford such an investment in salesforce, processing, and packaging equipment.

Farmers that cultivate species for a market where purchasers have never purchased before must create and develop the market. While developing a market for an unknown species is more difficult than overcoming unfavorable preconceptions about a species, it is a longer and sometimes more expensive process. Exporters of tilapia filets to the United States have successfully introduced a completely new species to the American seafood market (Watanabe, Losordo, Fitzsimmons, and Hanley, 2002). Price skimming and price penetration are possibilities with new items. If the marketing goal is to offer a product as a commodity in an industrial market, as an input into a larger supply chain, or to build specialized markets, the marketing plan should clearly state that. Despite this, niche markets are often thought of as low-volume, high-priced specialist markets, even though some experts claim that all markets are niche markets. Niche markets are often a small portion of a larger market. In specialized markets, sales are typically lower, but the aim is to sell fewer products at a premium price. Smaller businesses that uncover niche marketing opportunities may face less competition from larger corporations. A niche market is typically formed through a specialized contact, and the grower provides a custom product to that specific market. The grower will be responsible for providing all necessary materials for taste tests, sampling, and point-of-sale displays, as well as ensuring constant product quality. Direct sales from the grower to the end consumer have traditionally been the norm in aquaculture niche markets. As a result, the fish farmer performs supply chain responsibilities such as wholesaling, distribution, and retail. The grower keeps the profit margins from each of these phases in exchange. However, each of these responsibilities comes with a price tag, which can include the grower's time as well as fees for holding or processing facilities, utilities, labor, advertising, transportation, and packaging.

The chosen business structure must also be included in the marketing plan. Many fish farmers operate as sole proprietorships or partnerships, but others are vertically integrated businesses that control many stages of the market channel or supply chain. Vertically integrated is a shrimp company that has its own farm, hatchery, and packaging plant, for example. Other businesses merge horizontally to acquire market share by pooling marketing resources. Traditional lenders do not provide all of the funding for aquaculture enterprises. Some agricultural businesses also raise funds from outside investors. The majority of equity capital in agriculture has come through retained earnings, inheritances, and gifts. However, there are other ways to access equity capital. To pool funds and offer a source of equity capital, partnerships, joint ventures, and family corporations can be used. Outside investors seeking to conceal high non-farm income from taxes may be able to provide some equity capital. Limited partnerships have been employed in some agricultural companies, particularly cattle feed operations, to supply equity capital. However, as simply as it is made available, such a source of capital can be withdrawn. In such a situation, the investor assumes the risk of having limited options to sell their financial stake in the farm. Because additional stock must come from either present owners or new investors, it's critical to comprehend the costs. Current owners can give equity capital by reinvesting earnings or profits from the business or by putting money obtained from other sources into the business. Adding partners to the business can help secure outside equity money.

The cost of reinvesting retained earnings, also known as the necessary rate of return, is a sort of opportunity cost that compensates for the investment possibilities that are foregone. Furthermore, the opportunity cost is higher if the risk associated with the aquaculture business is higher than that of alternative investment opportunities. The expense of securing outside finance for the farm is proportional to the rate of return required to entice it. To attract investors, the rate of return from the aquaculture industry must be higher than that of other sorts of investment options. After taxes, equity capital is typically more expensive than loan capital. This is especially true when the cost of retained earnings is taken into account, as it should be in any thorough analysis. Varying lenders will have different preferences for different sorts of loan conditions. High-risk loans are frequently avoided since they erode credit quickly. To limit the overall extent of losses, high-risk loans often have limited lending limits. Lenders appreciate those with strong management skills. The possibility of timely loan repayment increases with highly trained and experienced management. Self-liquidating and asset-generating loans are preferred by lenders. Self-liquidating loans are ones that are repaid with the money generated by the equipment or other capital uses before the new piece of equipment wears out. These are regarded to be loans with a reduced risk. If the loan matures sooner than the income created by the asset purchased with the capital can be paid, revenue from other sources may be required to make the payments. Crop inputs like fingerlings and fertilizer normally mature when the food fish are sold, while loans for machinery and equipment may mature sooner than the number of years the equipment would be used. An asset-generating loan is one that is used to purchase capital goods like equipment, ponds, or land that raise the value of assets in the aquaculture industry. As the value of the company's assets rises, they can be utilized as collateral in the future. Loans that are both asset-generating and self-liquidating are preferred by lenders. Lenders prefer those that generate assets but are not self-liquidating, as well as those that are self-liquidating but do not generate assets. Lenders do not prefer self-liquidating or asset-generating loans, and they are far less likely to be granted.

The lender may desire a loan term that is shorter than the asset's expected economic life expectancy. The borrower, on the other hand, may want a lengthier loan term. It may be required to have an open dialog in order to get an acceptable loan agreement. Depending on the sort of charter under which they operate, banks are regulated in different ways and by different agencies. Individual state banking commissions regulate state-chartered banks, whereas the Federal Deposit Insurance Corporation regulates federally chartered banks. To ensure a safe and stable financial industry, regulatory agencies and the rules that control them were established. The amount of capital reserves, lending limitations, and deposit insurance that each bank must keep are all outlined in bank regulations. The relevant regulatory bodies have the authority to audit and examine banks to ensure that they are complying with all applicable regulations. Examiners are the people that conduct the audits. Bank examiners follow rules for capital adequacy, liquidity, and asset quality that they employ every time they evaluate a bank. To guarantee that the bank operates with appropriate liquidity, each bank is required to keep specific levels of reserves. While the number of reserves required varies depending on the bank's charter and applicable rules, required reserves are typically around 3% of transaction balances. When a loan becomes unviable and the lender is forced to declare bankruptcy, the bank risks losing a portion of the lent principle. Banks must cover this share of the loss by transferring reserves. The amount that the bank can lend is reduced when reserves are shifted. As a result, sustaining loan losses decreases the amount of business the bank may perform because reserves are depleted. Loan limitations for banks with national charters must be less than 15% of the bank's net value unless the loan is secured by readily marketable assets. State-chartered banks' loan limitations vary depending on state regulations.

The potential borrower bears the duty of proving that he or she will repay the loan with interest in a timely way. Individuals with a track record of timely loan repayment will generally find it easier to obtain credit, borrow large amounts of money, and possibly obtain more favorable loan terms. The character of the individual, demonstrated management competence, financial situation as evidenced by the balance sheet, repayment capacity, the purpose of the loan, and the quantity of collateral available to back the loan are all important considerations in creating and developing credit. These numerous factors are weighted differently by banks, but most banks evaluate all of them to some extent. Personal integrity is a crucial trait of a potential borrower. A good corporate reputation can assist you to persuade the lender that the anticipated values in the business loan proposal are correct. Aquaculture is a management-intensive industry that requires experienced leadership. A successful loan application requires clear evidence of management experience from either the owner or the appointed manager. Farmers with a track record will be able to demonstrate their management skills. The borrower's financial strength or weakness is shown on the balance sheet. The balance sheet is considered the most essential financial statement by some lenders because it displays the degree of leveraging in the business as well as the availability of collateral to secure additional loans. The financial status of the company also shows whether its net worth has increased over time. The management and financial condition of the company benefit from the company's steady expansion. The lender is primarily concerned with debt repayment. As a result, the borrower must show sufficient liquidity and a favorable financial position. A detailed cash flow budget will show the company's ability to repay any new loans that are suggested. Self-liquidating loans are typically easier to finance than other types of loans. This is because the loan can be repaid by selling the crop or animals that were used to finance the loan. Because capital assets are not frequently sold to generate revenue, loans for their acquisition may require additional security.

Buildings, equipment, and, on rare occasions, livestock inventories can be used as collateral. The loan should be granted on the basis of clear evidence that the loan installments can be made from agricultural revenue rather than collateral. Lenders, on the other hand, require collateral to cover the risk of unforeseen adverse conditions leading to loan default. Personal assets, such as a home, have been used as collateral in aquaculture. Credit reserves refer to additional financing that a farmer may likely get if needed. One or more lenders may be willing to lend additional funds to a corporation or individual borrower at any moment to help finance transactions. On farms that have not borrowed the maximum amount computed by the lender while establishing the credit worthiness of the borrower, credit reserves exist. The profitability of the firm, the individual's balance sheet, net worth, and assets available for use as collateral, as well as the individual's demonstrated management abilities, will all influence credit worthiness. The maximum amount that can be borrowed will change in accordance with national and international economic developments. Credit reserves are especially significant when the farm is facing difficult situations. Economic downturns may result in reduced farm prices and revenue, or conditions may arise that cause price hikes for crucial commodities like feed. During these times of economic hardship, the agricultural business' cash flow may deteriorate, and credit reserves may be used to boost operating capital borrowing. Credit reserves can also be utilized to refinance loans, postpone payments, and carry over or extend loan maturities. Credit reserves can help manage debt on the farm, especially during times of financial hardship.

3.7. CREDIT RESERVES

Credit reserves indicate that the farm is not fully utilizing its borrowing capacity. Credit is a good source of liquidity, and if retaining credit reserves results in lower-than-optimal liquidity levels, it may be prudent to boost liquidity by utilizing credit reserves and borrowing more cash. Maintaining merely a little amount of credit reserves, on the other hand, increases financial risk because there is little to fall back on during poor economic times. Furthermore, when a company is heavily in debt, lenders have more power over it. The lender's assessment of the profitability and risks associated with lending to an individual borrower is called credit worthiness of the individual borrower. The lender evaluates a borrower's credit worthiness based on the evidence supplied by the borrower to ensure that lending risks are minimal and debt servicing meets the conditions of the loan contract. In consumer finance, credit scoring is widely employed. To create a credit score, important criteria are determined and a weighting factor is assigned. To establish a credit score, the weighted averages of these criteria are added together. The predicted profitability of the business, projected liquidity of the business, solvency of the firm, collateral position, and loan repayment capacity are all common credit features. The credit score will alter over time as the business's values for each attribute change. The borrower's credit profile and score are influenced by the assets available for loan security, repayment, and income expectations, and other financial management activities.

There are several distinct types of cash flow issues that aquaculture businesses face. The timing of goods sales is a common source of liquidity issues (Engle and Stone, 2013). Seasonal aquaculture firms, in which harvests occur only once a year, may experience liquidity issues because revenue is not available throughout the year to cover expenses when they are due. Similarly, the timing of spending can lead to cash flow concerns. Borrowing and loan payments, leverage and credit use, and corporate expansion are all factors that contribute to liquidity issues. Furthermore, if sales, expenditures, and payment schedules are anticipated erroneously, farm planning might be impeded. This generates a visible cash flow challenge that must be factored into agricultural business strategy. For the time being, production costs must be met until sales create revenue. Cash flow issues can also be caused by the timing of spending. If a shrimp farm plans to acquire new aerators across the farm and makes the purchase at the production cycle (start), the investment will result in a significant cash outlay at the start of the cycle, before shrimp have been produced and sold. As a result, during that time of year, there is no cash revenue to match the cash outlay. Because the surplus money was spent on the new aerators, if shrimp prices fall or feed prices rise, the farm may no longer have a surplus from the prior year to handle these adverse situations.

Cash flow and liquidity issues might arise as a result of business expansion. For example, the farm may opt to build a hatchery since meticulous study has revealed that vertically integrating the food fish grow-out business with a hatchery and fingerling enterprise will yield higher revenues (Hossain, Khan, Saha, and Dey, 2022). The hatchery will require a new structure, new ponds, and the creation of spawning broodstock. Selecting broodstock from existing fish stocks may reduce the weight of fish available for sale. The area of water available for food fish grow-out is reduced when ponds are used for fingerling production. The hatchery building and the ponds both require financial investments. A temporary monetary deficit may come from the combination of lower revenue from diverting fish to broodstock and reduced area, as well as cash outlays to construct the hatchery building. The monetary shortfall will vanish over time as new ponds are developed and fish stocks are rebuilt to previous levels, and the greater revenues will be realized.

CHAPTER 4

EXTENSIVE FISH FARMING

CONTENTS

4.1. Introduction 76

4.2. Strategies 76

4.3. Treatment 78

4.4. Combined Intensive-Extensive (CIE) System 79

4.5. Nitrification 81

4.6. Polyculture 83

4.7. Crabs 84

4.8. Shrimp 86

4.1. INTRODUCTION

According to a comparative economic analysis of the various prawn growing systems, MODEX or modified extensive systems becomes economically profitable only if the cost of production is reduced under current market conditions (Trino, Penaflorida, and Bolivar, 1992). Feed inputs, which account for 55% of production expenses, are options to save costs. When compared to the same diet with vitamin supplementation, removing the extra vitamins from the diet results in a feed cost savings. Vitamins are vital food components in the diet; however, they can also be found in ponds in the form of species. These food species are developed during pond preparation and maintained during the culture period under MODEX to provide a natural food source for the prawns. Natural food continues to make up a significant component of MODEX's total food sources for prawns. For the first several weeks, it provides a significant amount of protein and other nutrients that the prawns require. With increased prawn biomass, additional feed becomes necessary. Because supplemented feed is one of the most expensive inputs in the production process, cutting feed costs is critical to making MODEX economical.

4.2. STRATEGIES

One strategy to cut feed costs is to eliminate vitamin supplements from the diet. Some water-soluble vitamins are dispensable in purified diets. In intensive tilapia farming (Figure 4.1), it was determined that vitamin premix in the feed is not economically justifiable because it did not boost growth rates or food conversion. It was observed that removing vitamins from pelleted feeds had no effect on the growth of *Penaeus vannamei* (Figure 4.2) kept in pens for 28 days, implying that vitamin supplementation is unnecessary when natural foods are present in ponds (Chamberlain and Lawrence, 2009).

Figure 4.1. Intensive tilapia farming.

Source: https://aquaculture-israel.com/Technology/large_scale.JPG.

Figure 4.2. *Penaeus vannamei.*

Source: https://www.fao.org/fishery/docs/DOCUMENT/aquaculture/Cultured-Species/file/en/images/femalebroodstock.jpg.

Growth, survival, net output, and net cost of production per kg of *P. monodon* in the treatment with a diet that had no vitamin supplement were not significantly altered. In contrast, it was found that *P. monodon* on a no-vitamin diet grew the slowest. As a result, the lack of vitamins in the diet has a significant impact on prawn growth and survival. However, prawns and other animals have minimal vitamin requirements, and inadequacies in the diet could be addressed by a variety of alternative sources. For example, since the prawns were grown in tanks with filtered seawater, low growth performance on the no-vitamin diet could have been the result of relying completely on artificial feeds. Crustaceans, copepods, polychaetes, and nematodes were found to be essential natural food components of the pond ecology in experiments conducted. The same species, as well as chironomid larvae, were numerous in all ponds 5 days before and for 60 days after stocking; after that, their populations fell throughout the culture period, possibly due to prawn grazing (Hasan et al., 2011). The abundance of polychaetes, copepods, and nematodes increased shrimp production in fertilized and feed ponds. Many of the feeds used, which are possible sources of vitamins, may have compensated for the lack of extra vitamins in the diet. Vitamins, on the other hand, are heat-labile and light-sensitive, so it's likely that vitamins in these meals were destroyed or lost during processing and storage.

4.3. TREATMENT

The control treatment is by far the least expensive when the cost of manufacturing per kg is calculated. However, the product's mean weight will cause it to be rejected on the export market and valued lower on the domestic market. From the aspect of economic production, a diet without vitamin supplementation remains the optimum treatment. Although the differences in relative growth performance, survival, net production, and net cost of production/kg of prawns produced with diet are statistically insignificant, its lower cost of 18.02/kg product would make it more profitable to utilize in a modified extended culture system (Lee et al., 2022). High-density intensive ponds enable for regulated growth of a high fish biomass while maximizing the utilization of water, space, and manpower. Because these ponds are artificially aerated, oxygen levels should not be an issue. Excessive amounts of harmful inorganic nitrogen species, such as ammonium or nitrite, are a major concern in such ponds. Flushing the ponds with large amounts of freshwater is a frequent solution. This option requires availability of water source and results in dirty water being discharged into a receiving water body. Biofilters are another popular alternative for recycling water. Biofilters for a large-scale operation are expensive to purchase as well as to operate.

4.4. COMBINED INTENSIVE-EXTENSIVE (CIE) SYSTEM

A combined intensive-extensive (CIE) system is another option that combines intensive and extensive components. Water is pumped back into the intensive ponds after being emptied from the intensive ponds into the extensive unit. The large pond serves as a biofilter, with the added benefit of encouraging fish development. The resulting system is a hybrid of two unique water bodies: the intensive one is heavily loaded and constantly mixed, while the second is less heavily loaded, stratified, and oxygenated mainly during the day. Algal production is a substantial source of oxygen and, to a lesser extent, carbon, whereas it is a minor source in the intensively fed and aerated components. Ammonium levels in large ponds are mostly controlled by algal uptake. As a result of algal production, an increase in ammonia in the water is commonly seen. Nitrification occurs in continuously aerated and mixed intensive ponds, but not in traditional ponds. Denitrification potential is strong in earthen extensive ponds; consequently, a sequence of nitrification-denitrification conditions can be expected to provide an additional nitrogen disposal mechanism. The combination of the two systems may improve the overall system's ability to remove harmful inorganic nitrogen. To accomplish this, the system must be designed optimally. The operational degrees of freedom in the system include the relative sizes and loads of the intensive and extensive components, as well as the rate of water exchange.

During a four-month growing season, ammonium, nitrite, and nitrate concentrations were measured in the intensive ponds. The average nitrite and nitrate concentrations were 0.5 and 1.5 mg/L, respectively, with no discernible difference between the input and outflow. Furthermore, the levels of nitrate and nitrite did not increase with time. In contrast to research showing extensive nitrification in circulated ponds, both of these findings indicate that no considerable nitrification occurred in the ponds. The pond's ammonium contents ranged from 1.0 to 5 mg /L, which is high enough to sustain nitrification (Le, Fettig, and Meon, 2019). Nitrification potential was assessed in samples of water from both the extensive and intensive ponds to test the variables impeding nitrifying activity in the intensive ponds. In the laboratory, samples from the ponds were incubated with constant shaking. It can be shown that nitrification began only after a 24-hour lag period in all of the samples studied and increased dramatically after that. This gradual process indicates a requirement to adjust from the pond conditions to the optimal continuous aeration conditions offered in the laboratory test, which are comparable to those found in intensive ponds.

The variation between periods of abundant oxygen supply and those with anoxic conditions prevail is one of the reservoir's inherent features. The reservoir is stratified in terms of oxygen, and conditions near the sediment are always anaerobic. Furthermore, throughout the night, the entire reservoir volume is oxygen-limited. Nitrification cannot occur in these settings. It was observed that nitrifying bacteria do not disappear under such conditions. Nitrifying bacteria (Figure 4.3) can be found in anaerobic deep sediments, anaerobic hypolimnion of lakes, wastewater reservoirs, and anaerobic fish pond sediments. The bacteria are adapting to the anaerobic environment. It's possible that when nitrification conditions improve, these bacteria must adjust to the optimal set of circumstances. Water in the intensive aerated ponds had a hydraulic retention time of around 5 hours, which is less than the adaptation time required to begin an effective nitrification process (Rahimi, Modin, and Mijakovic, 2020). As a result, quick flushing could be the cause of the system's low nitrification. The ammonium concentration in the control ponds was the same as the water entering the reservoir. Because of the slower flushing and removal of excreted ammonium, the ammonium content in the intermediately flushed pond grew and attained a higher steady state. In that pond, water and bacteria stayed for roughly 18 hours, which was less than the nitrification lag period.

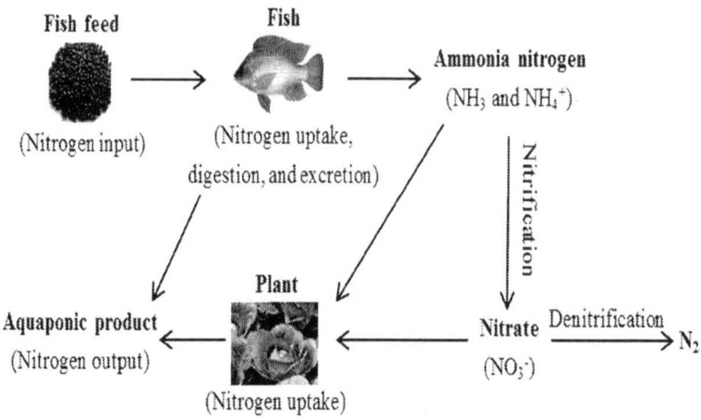

Figure 4.3. Nitrifying bacteria.

Source: http://www2.hawaii.edu/~khanal/aquaponics/nflow.jpg.

4.5. NITRIFICATION

Nitrification is another effective nitrogen removal method. Nitrification has two potential effects on hazardous inorganic nitrogen species. The first is ammonium and nitrite oxidation to non-toxic NO_3 species. Second, in the earthen extensive component's sediments, oxidized nitrogen can be rapidly denitrified and eliminated from the system as N_2. The presence of the intense component, which is characterized by ideal nitrification conditions, is not the essential requirement for successful nitrification. The microorganisms' residence time in that component must be long enough for them to adapt to the favorable conditions. This observation is significant in terms of nitrification in fish ponds. Even though the conditions for nitrification appear favorable, it was observed that the nitrifying bacteria are not present in an active form in the CIE system: the nitrification substrate, ammonium and a high oxygen level was found in all the intensive ponds and during part of the day in the reservoir. Nonetheless, it appears that even brief exposure to anoxic circumstances caused the bacteria to develop a physiological adaptation required to survive in anoxic environments. A significant lag period is required to recover from this state and return to an active nitrifying form. As a result, when water is delivered to optimal nitrification conditions in the intensive unit from the semi-aerobic reservoir, nitrification does not occur before the required adaptation time. When water samples from a non-flushed intensive pond were analyzed, nitrification occurs, with no lag time.

It was discovered that extending the hydraulic residence time in intensive ponds, i.e., slowing down the flushing rate, can result in efficient nitrification. Another alternative is to use water to slow down the bacterial discharge.

A biofilter (Figure 4.4) is a critical component for compensating for the short hydraulic residence time in heavily cleaned water. As a result, it can be demonstrated that in a combined system, two approaches can lead to effective nitrification. The first is to lower the water exchange rate, allowing for a hydraulic residence duration of more than 24 hours; the second is to add a biofilter to the system. The culture of Australian crayfish (Figure 4.5) of the species Cherax continues to pique people's curiosity. In Australia, three species, the marron, *C. tenuimanus*, the yabbie, *C. destructor*, and the red claw, *C. quadricarinatus*, are commercially produced, and research is undertaken to assess their potential (Doupé, Morgan, Gill, and Rowland, 2004). The yabbie is found throughout temperate and subtropical south-eastern and central Australia, where it is the subject of a long-running recreational and commercial fishery and is used in a variety of aquaculture endeavors.

Figure 4.4. Biofilter.

Source: https://www.innovationnewsnetwork.com/wp-content/uploads/2021/04/CERMEDI1-27480-image-1-696x392.jpg.

Figure 4.5. Australian crayfish.

Source: https://www.fao.org/fishery/services/storage/fs/fishery/images/culturedspecies//cherax/HarvestedRedclaw.jpg.

For the Cherax culture grow-out phase, various systems have been studied. Some researchers developed battery systems for intensive culture and evaluated growth under intense conditions in the laboratory, while semi-intensive culture in pools was tested. Most commercial plans for species utilize semi-intensive systems with hatchery stocking, pelleted feed, and intermittent aeration. *C. destructor*, on the other hand, is expected to thrive in large-scale aquaculture for the red swamp crawfish, *Procambarus chrkii*, in the southern United States. *C. destructor*, like *P. chrkii*, is a floodplain species that can withstand low oxygen, a wide range of temperatures and salinities, is polytropic, has a long breeding season, and can survive dry spells by burrowing. The extended grow out system consists of 0.2 ha of shallow, gently sloping ponds with the perimeters seeded with a crop of terrestrial plants. Every 6–12 months, the ponds are emptied and harvested, then left to dry for several weeks until a fresh crop is planted. The ponds are filled with yabbies, and the water level is gradually raised, drowning the crop while yabbies browse. Animals that reach market size are captured, while slower-growing individuals are transported to neighboring ponds where a crop has been prepared at drain-harvest.

4.6. POLYCULTURE

Polyculture, which brings together various species of fish living in diverse biotopes, benefits from the manure since it stimulates the entire food chain, including decomposers. Herbivorous, phytoplanktonophagous, zooplanktonophagous, molluscophagous, and detritivorous species, totaling 7 to 9 species, were introduced by Chinese culturists (Cvitanovic, Fox, and Bellwood, 2022). This production strategy produces a high yield but is difficult to control. The timing of manure application is determined by two key factors: rate of oxygenation as determined by fish behavior and algal density as determined by the Secchi disc technique or by plunging the hand into the water. In the event of anoxia or a high algal density, the environment is oxygenated and diluted with fresh, unfertilized water from lateral canals. Because the ponds are tiny, these activities can be undertaken. This type of aquaculture, which uses a huge variety of species and a lot of organic matter (OM), is quite complex and has been developed gradually and empirically. Many third-world countries would like to employ this system, but exporting it is difficult because it is impossible to transfer. The important step is to research the conditions under which aquaculture is integrated with other agricultural operations, establish which environmental characteristics are most relevant, and devise a method for objectively measuring them. Polyculture, which is necessary to maximize the system's value, can be accomplished by importing already-known species, but the diet of indigenous species can also be researched, with those that appear to be complementary being added to the association. Several scientific fields are involved in this strategy, but limnology is the one that should allow culturists to manage these artificial relationships.

4.7. CRABS

Mud crabs (Figure 4.6) are becoming a more important worldwide product in the Asia-Pacific region, where they are also the target of huge local fisheries. Mud crab species are a valuable source of revenue and fresh food for many coastal communities, and they are valued for their taste, texture, low fat content, and high protein, vitamin, and mineral content. For many years, mud crab production has been practiced in Southeast Asia, focusing mostly on the capture of immature crabs from the wild. Mud crab culture is the predominant species in Vietnam, and it is increasingly being seen as a viable alternative to shrimp culture, which has recently suffered from disease outbreaks, resulting in low and inconsistent yields. The mud crab, on the other hand, is a hard and dependable species that thrives in mangrove ponds. Extensive farming is the most prevalent culture method, while more intensive and cage culture methods are also used. A variety of problems are limiting the mud crab industry's potential growth in Vietnam. The first is the reliance on fishermen for supplies of natural crab seed. This fluctuating supply leads to erratic stocking schedules and densities. Mud crab hatchery production is expanding in Vietnam, resulting in a steady supply of crab lets for local farmers. Another restraint is the reliance on low-value fish diets, which are very expensive, subject to supply fluctuations, and can pollute the ecosystem. The development of a low-cost mud crab tailored food will help to considerably reduce the risk connected with mud crab farming in Vietnam.

Figure 4.6. Mud crabs.

Source: https://www.agrifarming.in/wp-content/uploads/2018/01/Mud-Crab.-e1523109342964.jpg.

On technical elements of mud crab grow-out aquaculture, there is a wealth of information available. However, few studies have looked into the economic sustainability of mud crab farming, particularly in Vietnam. Shrimp farming is one of the fastest increasing economic activities in Asia Pacific. Shrimp farming in Vietnam began around 100 years ago and was characterized by uncertain shrimp yields (Lan, 2013). Extensive shrimp farming systems began in the Mekong Delta of Vietnam shortly after the Vietnam War ended. Modern shrimp farming, on the other hand, began only after the country's economic reforms in the 1980s, when the government pushed shrimp farming. The government allowed saline and low-productivity rice fields to be converted into shrimp farms. Shrimp farming increased as a result. In Vietnam's Mekong Delta, the upgraded vast shrimp farming system is the most prevalent. Black tiger shrimp or *Penaeus monodon* are currently commonly farmed in the area. Although the criteria for shrimp farming system classifications vary, shrimp farming in the area can be divided into three categories: extensive, improved extensive, and intense.

4.8. SHRIMP

Despite significant advancements in shrimp farming, producers have operated their farms largely on their own experiences and information shared by their neighbors. Shrimp yields are poor and unstable in this system. While other Mekong Delta shrimp farming systems have been well documented, little is known about the improved extensive shrimp farming system. While the physicochemical state of the cultured ponds is important for shrimp growth, it is unclear whether the pond water and sediment in this system are optimal for shrimps. There is no general assessment of the ponds' physicochemical features. A vast range of physicochemical factors influence the survival and growth of black tiger shrimp. Low salinities are connected to shrimp infections, while salinities between 5 and 35 g L^{-1} are best for shrimp. Because it affects shrimp metabolism and feeding rates, water temperatures exceeding 32°C for lengthy periods of time might stress shrimp and limit growth. Dissolved oxygen (DO) levels should be higher than 4 mg L^{-1}, as lower levels can affect shrimp respiration and disease resistance. Total suspended solid (TSS) levels of 2 to 14 mg L^{-1} for black tiger shrimp or 50 mg L^{-1} for aquaculture in general are advised for black tiger shrimp (Suwoyo et al., 2020).

Shrimp can be stressed by low pH, resulting in fragile shells and poor survivability. Shrimp mortality is high at pH levels below 6. In most pond environments, nitrate is not poisonous. Total ammonium nitrogen (TAN) can affect shrimp at high pH levels because the poisonous gas NH_3 increases compared to NH_4–N. Even at low quantities, NH_3 can damage shrimp gills and decrease growth or induce death (Iber and Kasan, 2021). Shrimp are less hazardous to nitrite, a by-product of nitrification and nitrate reduction. Nitrite interacts with hemocyanin in shrimp blood at high amounts, reducing the blood's ability to transport oxygen. Hydrogen sulfide toxicity levels in shrimp are debated, ranging from non-detectable to 0.25 mg L^{-1}. In ponds without mechanical aeration, oxygen depletion is a risk when BOD surpasses 20 mg L^{-1}. Shrimp growth can be slowed by high Fe concentrations. Elevated OM collected on the pond bottom is the principal source of sediment impacts on shrimp. Low DO conditions come from sediment OM decomposition, resulting in lower harmful chemicals like nitrite, NH_3, and CH_4, which can affect farmed shrimp growth. Many upgraded vast shrimp producers were in debt at the end of 2008 due to shrimp losses (Nguyen, Vromant, Hung, and Hens, 2008).

The upgraded extensive shrimp ponds' unfavorable physicochemical features make them unsuitable for shrimp farming. Poor water and sediment quality can cause stress in cultivated shrimp, making them more vulnerable to disease. For improved shrimp production in the Mekong Delta of Vietnam, proper planning and adjustments of farming techniques are strongly advised. The following steps should be taken to improve the pond's physicochemical condition. It is important to allow sedimentation before shrimp culture ponds. We need to include no-culture breaks between shrimp crops. Drying and exposing the pond bottom to sunlight for a reasonable period of time after each production cycle is important. We need to remove all sediments produced after two production cycles and it is important to control the pond's vegetation to mitigate the condition of oxygen shortage. Research in this area must concentrate on developing a synchronous irrigation system for better water management within the dense river/canal system.

At the genus level, eutrophic water indicators like Peridinium and Pediastrum species were discovered. However, these species can also be found in mesotrophic conditions. At the end of the dry season, the former was common in the shrimp ponds. Cyclotella species, which are found in oligotrophic waters, were uncommon. In June, the largest concentration of blue-green algae indicated poor water quality. Blue-green algae were shown to be positively linked with H_2S. Phytoplankton species like *O. limosa, O. formosa, Anabaena sp.,* and *Phormidium tenue* that can create cyanotoxins such as microcystins, anatoxin-a, and homoanatoxin-a were found (Quiblier et al., 2013). Microcystins-LR and anatoxin-a LD50 values for animals have been published. Cyanotoxins have been found in aquatic vertebrates and invertebrates, including as mussels, and zooplankton. The peak of zooplankton coincided with the peak of phytoplankton in April. At the completion of the dry season, increased phytoplankton density most likely aided zooplankton development.

Except in August, when rotifers took over, copepods were the most dominant. Rotifer species that indicate eutrophic environments like *Brachionus falcatus* (Figure 4.7), *Brachionus angularis, Filinia terminalis* were only found infrequently. Copepod dominance in shrimp ponds has also been documented in Brazil. Cladocerans were the group with the fewest species. Protozoans made up a modest percentage of the total in April, but rose during the rainy season. *Copepoda nauplii*, a rich source of food for shrimp juveniles, were shown to be positively associated with salinity. Furthermore, no zooplankton species prevailed. On all sample days, all species made up 30% of the zooplankton population. Rotifers had the most species in August and October when compared to the other sampling days. In the ponds, Zoobenthos were scarce.

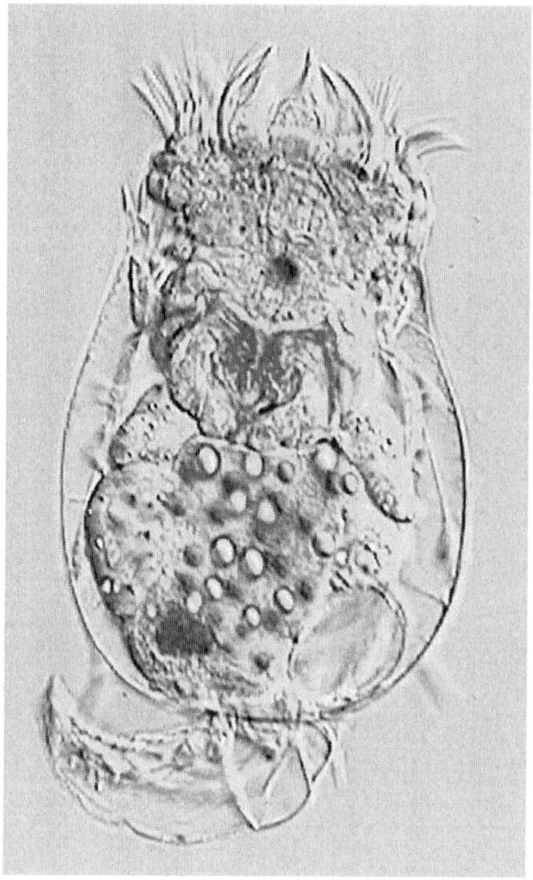

Figure 4.7. Brachionus.

Source: https://upload.wikimedia.org/wikipedia/commons/thumb/4/40/Brachionuscalyciflorus.jpg/800px-Brachionuscalyciflorus.jpg.

CHAPTER 5

BEST MANAGEMENT PRACTICES IN FISH FARMING

CONTENTS

5.1. Introduction 90
5.2. Aquaculture 91
5.3. Best Management Practices (BMPS) 94
5.4. Current Trends 95
5.5. Application 97
5.6. Waste Management 98
5.7. Restrictions 100
5.8. Environmental Problems 102
5.9. Initiatives 103
5.10. Development 105
5.11. Watershed Ponds 107
5.12. Brackish Water 112
5.13. Commercial Production 119

5.1. INTRODUCTION

World aquaculture production was over 5 million tons in 1980, accounting for about 7% of total global food supply. Aquaculture produced almost 46 million tons of fish and shellfish in 2004 accounting for nearly half of global food fish production. Since 1980, the yearly rate of rise in aquaculture production has been around 10%, which is quicker than any other animal food producing sector (Halvorson and Smolowitz, 2009). Global demand for fish will continue to rise due to population expansion, rising per capita incomes, urbanization, and a growing understanding of the importance of seafood in human health. Capture fisheries (Figure 5.1) must produce a significant portion of the world's fish and shellfish supply, but significant reforms are required to ensure that marine resources are managed sustainably. Seas must be protected from pollution and global climate change, and marine fisheries must be managed intelligently to recover and sustain the biodiversity and biological integrity of the oceans. Current scenarios for fisheries management, however, indicate that increasing marine fisheries landings beyond 1980s levels will be impossible. As a result, aquaculture must continue to expand in order to fulfill any increase in demand for fisheries products.

Figure 5.1. Capture fisheries.

Source: https://aquaculturemag.com/wp-content/uploads/2017/11/BL22_AQUA_2903350f.jpg.

5.2. AQUACULTURE

Aquaculture began thousands of years ago as a similar activity to other animal husbandry operations, with similar roots and purposes. In Asia, simple yet sophisticated fish rearing systems were devised that, in many ways, mimicked the fundamental ecosystem dynamics of natural aquatic systems. Aquaculture has been done as a relatively low-input activity for thousands of years, and it has been viewed as a useful business that provided high-quality animal protein for families or local communities.

Much of modern aquaculture is still based on these historical methods. Grass carp (Figure 5.2) or *Ctenopharyngodon idella*, silver carp (Figure 5.3) or *Hypophthalmichthys molotrix*, bighead carp (Figure 5.4) or *Aristichthys nobilis*, and common carp or *Cyprinus carpio* are the species raised in traditional pond aquaculture in China thousands of years ago. It accounts for approximately 90% of global aquaculture production. In reality, for local consumption, freshwater pond culture of cyprinids accounts for over a third of global aquaculture production, including plants. Culture methods and technologies emerged in response to profit motivations and support from aquaculture development organizations as aquaculture production grew quickly in the latter half of the 20th century. The purpose of supplying food for local consumption was sometimes replaced with the goal of generating higher-value products for export. The employment of culture practices with higher rates of resource use and larger environmental implications than traditional aquaculture methods was linked to this trend. The fast expansion of aquaculture happened at a time when environmental awareness was at an all-time high, and well-publicized in specific sectors prompted a deeper look at aquaculture in general.

Figure 5.2. Grass carp.

Source: https://upload.wikimedia.org/wikipedia/commons/thumb/b/b9/Ctenopharyngodon_idella_01_Pengo.jpg/800px-Ctenopharyngodon_idella_01_Pengo.jpg.

Figure 5.3. Silver carp.

Source: https://upload.wikimedia.org/wikipedia/commons/9/9f/Hypophthalmichthys_molitrix_Hungary.jpg.

Figure 5.4. Bighead carp.

Source: https://upload.wikimedia.org/wikipedia/commons/thumb/6/6c/Tolstolobec_pestr%C3%BD.jpg/800px-Tolstolobec_pestr%C3%BD.jpg.

Aquaculture production may and must rise in the foreseeable future, but any expansion must be done responsibly. Consumer food preferences and the capacity of local, regional, and global ecosystems to support aquaculture production are just a few of the difficult and sometimes conflicting issues and considerations that must be balanced during development. Some problems will be tough to resolve. A major critique of aquaculture, for example, is the growing of high-value, carnivorous species that require significant ecological support to produce. Many of the environmental issues attributed to aquaculture, on the other hand, can be solved by changing or enhancing culture methods. Environmental implications must be detected, and the aquaculture community must respond by modifying practices or inventing new technology to prevent negative effects. Aquaculture has achieved significant progress in terms of resource efficiency and environmental performance, and more progress is possible.

The four deep-south states of Alabama, Arkansas, Louisiana, and Mississippi account for over 60% of aquaculture production value in the United States ("History, Status, and Future of Aquaculture in the United States," n.d.). The predominant species raised in Alabama and Mississippi is channel catfish, whereas catfish and baitfish are raised in Arkansas. Crawfish, oysters, alligators, catfish, pet turtles, and baitfish are all important products of Louisiana's aquaculture business. Idaho, Washington, California, and North Carolina are the top producers of trout in the United States. The most important crop in Florida's aquaculture business is freshwater ornamental fish, which generated sales of more than $40 million in 2005 (Shelton and Rothbard, 2006). Aquaculture ponds are enclosed areas of standing water used to raise finfish or shellfish. Ponds can hold fresh, brackish, or saltwater and are normally made of soil, however some are coated with plastic or other materials to prevent seepage. The long hydraulic residence period, which permits much of the waste created during cultivation to be eliminated before water is drained, is an important aspect of the ponds. Ponds are similar to water recirculating aquaculture systems (RAS) in this way, except water in recirculating systems is treated in separate unit operations, whereas waste treatment and aeration are built into the pond ecology. In 2003, ponds around the world produced almost 27 million tons of fish and crustaceans, accounting for roughly 70% of global aquaculture production. More than 310,000 metric tons of fish, crawfish, and marine shrimp were produced on 140,000 ha of ponds in the United States in the same year, accounting for nearly 65% of US aquaculture production (Tidwell and Allan, 2012). Inland, freshwater ponds mostly used to cultivate channel catfish, crawfish, bait and ornamental fish, and hybrid striped bass produced about 90% of the finfish produced in the United States. Marine shrimp are raised in a few 100 hectares of ponds along the Texas coast.

5.3. BEST MANAGEMENT PRACTICES (BMPS)

Best management practices or BMPs are becoming more widely recognized as important targets for reducing on-farm and processing impacts, as well as the overall impact of agriculture. BMP adoption and encouragement are frequently considered as a means to an end in the pursuit of continual improvement in environmental performance. If best practices are in place, the presumption is that producers are doing everything possible to avoid the worst effects of production. In actuality, producers' environmental performance will vary depending on whatever BMP they use, with the majority falling somewhere in the middle. Even while there can only be one optimal practice, many others are likely to be better. Furthermore, today's best practice will become the usual tomorrow and, at some point in the future, the worst practice, to be avoided. If gradual improvement is the goal, tomorrow's BMPs will almost certainly be better than todays. As a result, the term BMPs are rapidly being phased out in favor of better management practices. The acronym BMP applies to both. The distinction indicates a fundamental shift in thinking.

One change is the realization that no single BMP decreases a critical impact in the same way for all producers in a given country or all producers of a specific species internationally. There isn't a solution that fits everyone. The most successful BMPs are determined by a variety of factors, including the species being cultivated, the nature and magnitude of the impact, the volume of production, the producer's resources like money, labor, and land, and the overall management system already in place. BMPs are frequently used as proxies by governments and certification programs since it is easier to evaluate whether a practice is in use than whether an impact has been significantly decreased. However, meaningful proxies should be derived from data, not used in place of it. Furthermore, many producers prefer to know what to do and not have to think about it. BMPs, on the other hand, are an inexact shortcut in practice. It's more difficult to think about how to achieve a performance result. Prescriptive solutions are more likely to result in compliance, but they might also fail. The accomplishment of meaningful outcomes is contingent on the BMP's ability to considerably lower a specific impact, as well as if the BMP chosen was the correct one and was successfully executed by the producer.

5.4. CURRENT TRENDS

According to current trends, future results will be more targeted and measurable, with less focus on how they are attained. BMPs have been confined to guide documents and workbooks, while they were once an explicit element of rules and standards, whether compulsory or voluntary. As a result, the global focus on BMPs is turning away from prescription and toward assisting producers in improving performance against a baseline while also allowing them to develop their own methods of improvement. In a world where lower impacts are sought by governments, investors, customers, and non-governmental organizations, we need to examine the opportunities and limitations of BMPs as a tool for environmental management, increasing overall environmental performance, and decreasing waste. BMPs are also a tool for producers to boost their competitiveness in increasingly competitive marketplaces. Producers originated the vast majority of BMPs, not researchers, regulators, government extension agents, buyers, investors, or non-governmental organizations. It goes without saying that BMPs exist for practically every known production activity, just as there are better and worse ways to do virtually every known production activity. BMPs are most easily marketed when they boost revenue immediately or in the near future. BMPs can be pushed as market rivalry grows because they cut costs and input utilization while also increasing overall efficiency. BMPs can be utilized to reduce fines, liability, and risk if producers are penalized for waste or pollution. BMPs have always been easiest to spread when they have a direct impact on revenue. Producers are less likely to accept BMPs that minimize effects but have no beneficial impact on income. However, when governments or the private sector propose or mandate BMPs, this can be the case. BMPs have been viewed as a way to improve overall environmental management and performance in recent decades. Nonetheless, there is a lot of overlap between the two goals: lowering inputs and improving resource efficiency per unit of production increases overall environmental performance. Inefficient resource usage can lead to pollution, among other things.

In agriculture, as global economic competition intensifies, the most efficient farmers will thrive. Their ability to innovate, recognize, or adapt strategies that reduce input use, waste, and pollution will define them. These manufacturers will be more profitable or, at the very least, remain competitive in the face of price deteriorations. Those that stay competitive will not just concentrate on improving a particular commodity. They will concentrate on their whole production methods and analyze what crops to grow on a regular basis in order to maximize physical, financial, and market advantages. Specific crops, as well as the methods used to grow them, will evolve. This would tend to prioritize BMPs that provide the best financial returns over those that may dramatically minimize important consequences. Agriculture experience suggests that there are BMPs for larger and smaller producers, as well as those with more and less money to invest. BMPs for agriculture, like other production systems, vary by location, crop, scale, and intensity of production. As a result, no single BMP or set of BMPs will work for all producers in the same way. Indeed, any BMPs that are advocated are likely to favor some producers over others. Because of their importance in reducing overall impacts and ensuring agricultural sustainability, some BMPs cannot be left to the market alone. Most producers around the world are unlikely to make the switch to better methods without assistance. In the short run, targeted government subsidies can offer incentives for the adoption of BMPs that address external factors that are unaffected by resource efficiency surges. BMP identification and adoption can also be aided by government regulatory procedures.

BMPs have played a similar function in aquaculture as they do in other industries. Governments, development agencies, input suppliers, industry groups, academics, and non-governmental organizations have promoted them as a way to cut costs and waste, increase income, reduce pollution, produce higher-quality products, gain, or maintain access to new markets, and get regulatory relief (Giri, 2017). BMPs are assumed to yield good, verifiable, and predictable outcomes on the ground. The effectiveness of BMPs in decreasing key impacts is rarely measured, making it far less evident than if they have been implemented. Most BMP-based programs focus on compliance rather than performance. Where there is a perceived benefit to the producer especially but not necessarily in terms of cash, where BMPs are a condition of allowing and licensing, or where adoption has allowed producers to bypass rules, BMP adoption has been most rapid. BMPs may or may not make financial sense for a variety of different-performing producers when they are mandated by law. There will be a need to determine the precise sorts of producers, performance levels, or cumulative consequences that necessary BMPs may not address as effectively as the most effective BMPs. BMPs are also more easily implemented if they improve a producer's operations. In fact, if BMPs contradict local beliefs, social acceptance may actually hinder their implementation. It was considered that building shrimp ponds in old mangroves was a good idea because that's where shrimp were found in the wild, and that building ponds above the high-tide mark was uneconomical (de Lacerda et al., 2021). Salmon should be fed high amounts of fishmeal and oil since that is what they would consume in the wild, according to retailers in the United Kingdom. Because they were based on inappropriate thinking, these beliefs hampered the adoption of improved methods.

5.5. APPLICATION

BMPs can assist manufacturers in increasing their efficiency. This equates to lower input consumption and costs. Less water exchange translates to less pumping, less energy consumption, and cheaper production costs. Improved feed conversion translates to less feed utilized, better water quality, less pollution, more profitability, better relations with neighbors and other resource users, fewer court challenges and legal battles, and a more secure operating license. More stocked animals are harvested, higher feed conversion, less downtime, less quarantine, less problems with neighbors, and higher earnings result from reduced diseases. Fewer escapes result in less money being wasted on animals that are never sold, faster feed conversion, fewer conflicts with other animals, and more earnings. In other words, any BMP that increases performance tends to lessen some, if not all, of the risks that come with it. However, there are a few crucial situations where the combined advantages of all BMPs are larger than the sum of their parts. Perhaps more crucially, producers can choose from a variety of BMPs to attain a comparable level of performance. As a result, each BMP performs differently in the hands of different producers, and there is overlap in performance amongst BMPs. Most producers achieve acceptable performance for some BMPs; however, only exceptional producers achieve acceptable performance for other BMPs. Producers can choose from a variety of management strategies for water and water quality, feeds and feeding, diseases, and escapes. Producers will accomplish various levels of environmental performance in terms of effluent pollution, invasive species, disease transmission, and genetic contamination by doing so.

5.6. WASTE MANAGEMENT

Waste (Figure 5.5) is now defined as anything that isn't put to good use. The majority of efficiency gains will result in waste reduction. Aquaculture facilities can generate physical waste or run inefficiently. Cost, rather than legislation, is one of the most powerful motivators for waste and pollution reduction. Cost is perhaps the most essential consideration for a producer. In business, input prices have a big influence on how people think about waste. It is more inefficient to not utilize an input efficiently the more expensive it is. It may be necessary to define acceptable and unacceptable performance levels from a society standpoint. Market techniques are used to implement voluntary standards. However, they are only optional to the extent that manufacturers desire to sell into different markets. Permits and licenses are used by governments to control and set additional restrictions. Scarcity awareness is a critical aspect in attributing value to an input when defining these performance levels. Although there is a widespread and increasingly reinforced belief that natural resources are finite, the attention has so far been exclusively on the environment and natural resources deemed valuable. Most producers do not consider wasting labor or an employee's skills. Even while biodiversity, natural habitat, and ecosystem services have intrinsic value, they are not seen as having market value. This attitude is shifting, and it will have a significant impact on how waste and inefficiency are defined. Changes in science, public attitudes, market pressures, or a combination of all three move the focus of waste or inefficiency. Water, carbon, energy, and greenhouse gas concerns will very certainly influence future standard development.

Figure 5.5. Waste.

Source: https://www.globalseafood.org/wp-content/uploads/2020/02/DAUDA-Pt-2-Pic-0_1500-960x720.jpg.

The majority of BMPs are created by producers in response to a specific problem like disease, higher input costs, laws, market needs, etc. BMPs were once thought of as tools that individual farmers might use or not use. Producer organizations are increasing, however, that the reputation of many can be harmed by the actions of a few. This can have a negative impact on local reputations, market access, and government regulation. As a result, BMPs are now included in many producer standards of conduct. Many manufacturers are hesitant to reveal financial data for fear of it being used against them or by competitors to obtain a competitive edge. Most producers want to increase production and net profits. Most governments, on the other hand, devote their limited resources to discovering BMPs that contribute significantly to the reduction of specific impacts. The United States imports around 85% of its fishery goods, with aquaculture-derived fish, shrimp, and molluscan shellfish accounting for a significant portion of this trade. In the 1990s, a number of environmental NGOs became critics of aquaculture, notably of marine shrimp and salmon farming (Zajicek, Corbin, Belle, and Rheault, 2021). Attempts by more environmental NGOs in the United States and other Western countries to enact consumer boycotts of these species were unsuccessful. Several moderate NGOs, on the other hand, continued to work to promote environmental and social responsibility in aquaculture. Aquaculture organizations have promoted environmental responsibility, governments have increased environmental regulations for aquaculture, and some retailers and other food service buyers are seeking environmentally and socially responsible aquaculture products as a result of criticism of aquaculture practices by the World Wildlife Fund, Environmental Defense, Conservation International, and other environmental NGOs. The majority of aquaculture's negative environmental and social impacts occur at the farm level, and producers have been encouraged to use BMPs to mitigate these effects. The United Nations Food and Agricultural Organization (FAO), the World Bank, and other international development agencies have also agreed to join the NGOs' efforts to promote responsible aquaculture. Furthermore, NGOs' training efforts to urge customers to make more responsible seafood choices appear to be having an impact on the market.

5.7. RESTRICTIONS

For years, numerous federal and state environmental restrictions have applied to aquaculture in the United States. However, a federal rule for aquaculture effluents was just recently developed by the United States

Environmental Protection Agency (USEPA). Environmental NGOs were instrumental in starting this regulatory process. Environmental NGOs tend to be interested in US aquaculture mostly as a result of their previous experiences with salmon and shrimp production in other countries. Domestic aquaculture is not a large source of pollution, according to the USEPA. The effluent rule essentially encourages operations large enough to be classified as concentrated aquatic animal production facilities to implement BMPs. Major purchasers seeking products from aquaculture will likely have a greater impact on US aquaculture than the USEPA effluent rule and other environmental restrictions. However, aquaculture in the United States and other wealthy countries differs significantly from aquaculture in developing countries. Most seafood is imported from developing countries, and aquaculture is limited to a few species of moderate to high value for domestic markets in wealthy countries. Channel catfish, rainbow trout, and crawfish raised in aquaculture in the United States, for example, are virtually entirely sold domestically. Aquaculture is used to produce trout, salmon, and bivalve shellfish in several European Union (EU) countries. Although some of this output is exported, the majority of it is destined for the European market. Carp production in Europe is primarily for local use. There is a wider range of culture species in developing countries with major aquaculture sectors than in affluent countries. Although the majority of the products are consumed in the country, several emerging countries export substantial quantities of aquacultured shrimp, tilapia, salmon, and other species. To be viable, farms in affluent countries must be fairly large, and there are relatively few growers. Because labor is expensive, mechanization, and automation are used to the greatest extent possible. To be viable, even small family farms must invest in cutting-edge technologies. Many developing countries have tens of thousands of aquaculture farms. Farms with water surface areas of 0.25 to 5 hectares are common in pond culture. Other industrial systems are also far smaller than those found in wealthy countries. Although there are some very big shrimp, tilapia, and salmon farms in developing countries, tiny farms in many countries, particularly in Asia, produce significantly more than huge farms. The majority of small-scale farmers are poor and live in rural areas. They are uneducated, uninformed about aquaculture technology, and unconcerned about environmental issues. Production procedures are frequently not monitored, and few producers employ BMPs to mitigate negative environmental effects. Large farms are more likely than small farms to employ highly qualified managers and employ superior production procedures. Large farms in underdeveloped countries are not as mechanized as those in developed countries due to the low cost of labor. Almost all major farms in emerging nations are geared toward export.

5.8. ENVIRONMENTAL PROBLEMS

The environmental consequences of dense clusters of farms are clear, and the risk of negative consequences is often worsened by a lack of efficient government regulation of effluents, wetlands, predator control, water consumption, and other variables. The environmental impact of several small farms can be as enormous as or larger than that of a large one. In industrialized countries, aquaculture facilities are often concentrated in specific places. Almost all catfish farms in the United States are concentrated in a few counties in Alabama, Arkansas, Louisiana, and Mississippi. About half of Alabama's catfish farms are located in a single catchment, yet the farms only cover 7.5% of the watershed's surface area. In general, developed countries have more fair social conditions than underdeveloped countries. Agricultural employees in many poor nations may be paid less than the legal minimum wage, working, and living conditions on farms may be bad, and there may be no benefits or job security. Large aquaculture firms may be given priority over other stakeholders when it comes to land, water, and other resources, causing problems with local populations. Companies may choose to hire laborers from distant locations over local workers. Some farms may use child labor as well. Small-scale producers rarely recruit labor, although they do expect under-age family members to help on farms. This is fine as long as youngsters are not denied access to education. Aquaculture clearly causes more social inequality in underdeveloped countries than in the United States or other rich countries. Environmental non-governmental organizations continue to oppose aquaculture for environmental and socioeconomic reasons, but talks with leaders from several of these groups indicate a shift in perspective. They've realized that aquaculture is an important part of the global food supply system, and that when done correctly, it may be more environmentally friendly than fishing. Aquaculture is a type of agriculture, while fishing is similar to hunting and gathering. Hunting and gathering are no longer the primary sources of food and fiber from terrestrial sources for humans. Similarly, we should increasingly look to responsible aquaculture as a source of aquatic food rather than fishing.

5.9. INITIATIVES

Retailer initiatives can help provide safe, high-quality items while also minimizing harmful environmental and social consequences. The standards, on the other hand, are tailored to the marketing goals and consumer base of the shops. The standards may not be made public, and the process of developing standards is not necessarily open. Professionals with experience in aquaculture and environmental issues are being hired by several shops. Others are hiring environmental non-profits to help them set purchasing strategies and standards. Of course, this strategy might also be used to gain the NGO's approval. The fundamental goal of retailer programs is for a company to distinguish apart by creating its own set of standards. This indicates that the retailer is aware of and in control of the suppliers. The requirements also show that the shop is looking for a product tailored to its customers. These programs are intended to satisfy a company's client base while also demonstrating that the corporation cares about the environment and social issues. Better management practices for aquaculture are intended to create an on-farm management system that minimizes negative environmental and social impacts while also ensuring food safety. Since the late 1990s, public agencies and farm organizations in the United States have pushed BMPs as an alternative to prescriptive regulations for achieving environmental protection or conservation goals (Liu, Bruins, and Heberling, 2018). Despite the fact that the motivation for developing BMPs has ranged from selfless self-governance to governmental mandate, the process of developing, implementing, or verifying BMPs has many similarities.

Producers, input suppliers, fish processors, seafood buyers, consumers, government agencies, residents of affected communities, and environmental, social, and consumer activists are all touched by aquaculture. Each group's representatives should be involved in the development of BMPs. It is critical to obtain agreement among stakeholders on the challenges that BMPs will address. Some stakeholder concerns may appear insignificant to scientists and other technical experts, but all concerns must be openly discussed in order to create a credible and transparent process that will result in an effective set of practices that will be acceptable to all parties involved. Once the causes of concern have been agreed upon, the stakeholder group should appoint technical experts to create BMPs that address each issue. Naturally, all stakeholders, particularly producers, should be involved in the revision of BMPs. All significant issues must be addressed, but the BMPs must also be realistic, implementable, effective, and acceptable to the stakeholders. Of course, the competent agency must verify compliance with government-mandated BMPs. BMP adoption in a certification scheme necessitates third-party verification. This procedure should be considered as continuous. BMPs can be changed or new BMPs produced for implementation based on the results of the monitoring program to verify performance.

BMP-based programs provide significant flexibility in achieving environmental protection goals. The BMP approach does not prescribe specific procedures, but rather provides a range of possibilities from which producers can choose and adapt to site-specific requirements. BMP initiatives might be voluntary or mandated to implement. Environmental protection is the goal of voluntary programs, which also increase market prospects for companies. Governments utilize mandatory programs instead of prescriptive numerical restrictions to achieve environmental protection. The most desirable type of BMP implementation is voluntary adoption, which indicates that producers recognize the need for responsible aquaculture and are prepared to put in the work and pay the costs to achieve it. In actuality, however, voluntary BMP adoption is perhaps the riskiest way of deployment. Producers might evaluate their operations and simply state that they are already following the best practices. The Global Aquaculture Alliance or GAA, for example, produced a set of BMPs for shrimp farms to follow on a voluntary basis. Commitment to BMP compliance would show environmental and social stewardship, which would appeal to hesitant investors or lenders. Studies demonstrating the cost effectiveness of BMPs would be particularly useful to those considering aquaculture facility investments or loans. Large aquaculture projects in underdeveloped nations are frequently financed by the International Finance Corporation or IFC. The IFC requires its clients to follow certain environmental regulations. Adoption of BMPs may become a standard condition for farmers seeking investors or loans in the future.

It is necessary to publish research findings and extension recommendations on feed use efficiency, aeration, effluents, and other areas of production with possible environmental consequences. Water usage within the culture facilities must be meticulously documented. Wasteful and environmentally hazardous aquacultural activities should be recognized after careful consideration of resource availability, competing applications, and potential environmental implications. Furthermore, the potential for damaging elements of farms should be investigated. Investigators should compile a list of BMPs that could be effective in avoiding resource waste and other negative environmental and social consequences. BMPs should also include farm-level actions for preventing contamination of aquaculture products. Reports on the environmental state of aquaculture sectors, such as those for marine shrimp farming, are available. Producers, local extension specialists, local environmentalists, regulatory authorities, and representatives of communities located near farms or facilities should all be included among these stakeholders (Nesheim et al., 2015). The team in charge of the environmental survey and report, as well as aquaculture experts, representatives from international NGOs, officials from processing factories, and maybe others, should be included. Environmentalists and industry officials have typically had the most heated discussions over aquaculture's environmental and social implications during multi-stakeholder meetings. These conflicts are marked by passionate arguments about the types and severity of aquaculture's environmental implications, leading to extreme viewpoints and little compromise. It's worth noting that some of the firmest opponents of environmentalist claims are attempting to address environmental problems through BMPs and norms of conduct within the aquaculture sector. As a result, the aquaculture community's and environmentalists' interests are aligned, and it makes sense to include environmentalists more explicitly in BMP creation procedures.

5.10. DEVELOPMENT

For initial BMP development, a smaller subset of the primary stakeholder group can function as a technical advisory committee in some situations. Because many stakeholders have a vested interest in the BMPs being established, restricting participation in stakeholder meetings is challenging. As a result, meeting organizers and facilitators should have a good general awareness of the people and groups who would be most affected and interested in the sort of aquaculture being addressed. Consultants and professional facilitators are frequently employed to perform this task. Nonetheless, the problems should be thoroughly discussed because the exclusion of a concerned party could jeopardize the BMP program's legitimacy in the future. Ponds are where the majority of the world's fish and crustaceans are raised. Ponds are utilized to raise channel catfish or *Ictalurus punctatus* and crayfish or *Procambarus* spp. in the United States, which account for about 80% of the country's annual aquaculture production by weight. Baitfish, hybrid striped bass, freshwater ornamental fish, and penaeid shrimp are all commercially important crops grown in ponds. In the United States, around 125,000 ha of ponds are used for aquaculture. Aquaculture ponds are limited basins of standing water that are managed to generate a crop of the aquatic species of interest. To reduce seepage, some ponds, for example, may be lined with plastic. Suitable water quality is maintained in recirculating systems by discrete unit water treatment operations and continuous water exchange in flow-through and net-pen systems.

Although all aquaculture ponds appear to be similar on the surface, their environmental implications and sustainability can differ significantly. Culture intensity, pond hydrology type, and the species grown are all factors that affect the environmental performance of pond culture systems. Extensive pond culture systems have limited water exchange, rely mostly on the generation of natural feeds to maintain culture species growth, and have low crop yields as a result. Intensive pond systems, on the other hand, may have higher water exchange rates, rely on manufactured feeds to encourage rapid fish or crustacean growth, and produce relatively high crop yields. Semi-intensive systems are in the middle of the intensity spectrum. The link between culture intensity and long-term viability is indirect. On the surface, widespread aquaculture appears to be more sustainable. However, metrics of resource usage efficiency in production such as land, water, fossil fuel energy, and nutrient inputs are significant indicators of environmental sustainability, and efficiency varies by culture system and index. For example, intensive systems produce less per unit fossil fuel energy input than extensive systems, whereas extensive systems produce substantially less per unit land area than intensive systems. While combining the different efficiency criteria used to estimate sustainability is complex, middle levels of culture intensity are regarded to be the most sustainable. Environmental sustainability should be assessed using a variety of criteria rather than a single set of operational characteristics. Furthermore, sustainability encompasses the economic, social, and environmental spheres. In the United States, extensive pond aquaculture systems are not economically viable, and all commercial pond aquaculture is based on either semi-intensive or intense culture of commodity goods or intensive culture of high-value products.

The classification of aquaculture ponds by hydrological type is significant because pond hydrology affects water use, effluent volume, and, to a lesser extent, effluent quality. Pond hydrology has an impact on effluent management strategies and the ability to control effluents. Embankment ponds (Figure 5.6) are the most popular type of aquaculture pond, and they are built by grinding material from the pond bottom and utilizing it to build embankments around the pond perimeter. Alternatively, for the majority of the pond area, the original bottom elevation is preserved, and the soil needed to build embankments is retrieved from a trench constructed adjacent to the embankment. Because catchment areas are limited, consisting solely of the pond surface and interior embankment slopes, pumped water is required to fill ponds and keep water levels stable during droughts. In humid climates, precipitation that falls directly into a pond can be a significant source of water. Direct precipitation raises the pond water level to the same level as rainfall. The frequency distribution of precipitation events is roughly log-normal: there are significantly more low-amount precipitation occurrences than high-amount precipitation events. Seasonal variability and rainfall extremes are concealed by long-term average rainfall. The variation in monthly rainfall between years is large, especially throughout the winter and early spring. In the summer, rainfall is more variable than in the spring. The rainy and dry seasons are distinct in most tropical places, which have a significant impact on pond water budgets. Providing storage volume in ponds to catch rain reduces the requirement for pumping water, especially during the summer, and reduces the amount of effluent produced after pond overflow. Many farmers make an effort to capture this free source of water.

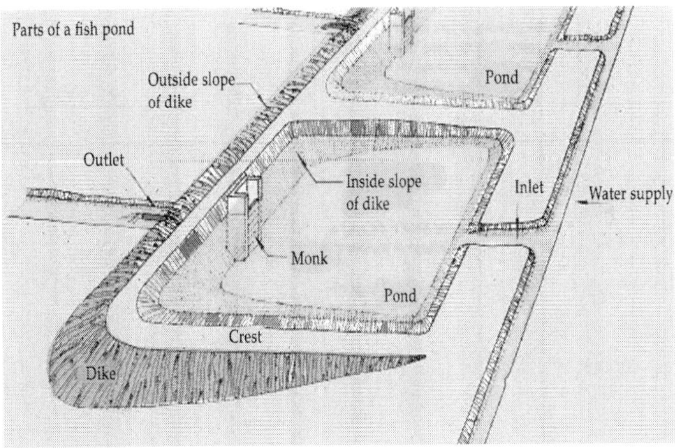

Figure 5.6. Embankment ponds.

Source: https://www.fao.org/fishery/docs/CDrom/FAO_Training/FAO_Training/General/x6708e/GR000000.JPG.

5.11. WATERSHED PONDS

Watershed ponds (Figure 5.7) are built to catch overland flow from the surrounding watershed, particularly if groundwater or surface water supplies are unpredictable (Eshleman, Pollard, and Kuebler, 1992). The volume of runoff is determined by watershed factors such as slope, vegetation cover type and extent, soil type, and moisture. The volume of runoff is also affected by the amount and severity of rainfall. Many ponds in west Alabama and east Mississippi are hybrids of watershed and embankment ponds, and many catfish farms in the area include wells to augment runoff inflow. Because runoff is a far more important supply of water for watershed ponds than levee ponds, water level fluctuations in watershed ponds will be greater than in levee ponds. During droughts, water levels in ponds that rely only on runoff can drop dangerously low. Groundwater has both good and unfavorable features for use in aquaculture ponds, but it is an ideal source of water for pond aquaculture on the whole, as it is usually free of chemical and biological contaminants found in surface waters. Groundwater is rarely susceptible to the changes as surface fluids. Groundwater is typically cooler than surface waters, lacks dissolved oxygen (DO), and may contain higher quantities of reduced chemicals that are harmful to fish. Aquifers are underground water reservoirs that are frequently associated with previous alluvial soil layers. Shallow depth to aquifer surface, high transmissivity and hydraulic conductivity, abundant alkalinity and hardness, separation from potential surface contamination by an impervious layer, and recharge that maintains aquifer volume are all desirable characteristics of groundwater aquifers that are suitable as a water supply for aquaculture ponds.

Figure 5.7. Watershed ponds.

Source: https://gripp.iwmi.org/wp-content/uploads/sites/2/2018/08/Nature-based-integrated-watershed-management-solutions-4-768x534.jpg.

To complement groundwater and runoff sources, water can be piped into ponds from streams and rivers. Surface waters, on the other hand, are unattractive from many perspectives. The seasonal availability of surface waters does not always correlate to pond aquaculture demand. Although peak stream discharge occurs in the southern United States from late winter to early spring, demand for surface water, particularly as a supplemental water supply, is highest during the summer, when stream discharge is near the annual lowest. Surface water availability varies greatly from season to season and year to year. In severely dry years, there may not be enough surface water to augment other water sources. Flooding of streams and rivers during exceptionally wet years may cause damage to pond levees and drainage facilities. Floodwaters may overtop levees, allowing cultured stock to escape. Surface waterways are particularly problematic since they can be contaminated by wild fish, disease infections, agricultural pesticides, fertilizers, and silt. Surface waters in the United States are classified state waters, and hence may be subject to regulatory restrictions that limit withdrawals for pond aquaculture. Evaporation, seepage, and overflow occur when inputs exceed pond water storage capacity, resulting in water loss from ponds. Because this water is not available for downstream applications, the sum of evaporation and seepage reflects consumptive water consumption. Evaporation from the pond surface to the atmosphere is known as pond evaporation. The rate of evaporation rises with wind speed and the difference in vapor pressure between water and air, and falls as the latent heat of vaporization and water density rise. Evaporation rates in the temperate zone tend to follow annual temperature cycles, so they are highest in the summer and lowest in the winter. Pond evaporation is difficult to quantify directly in practice because changes in pond water are due to the shared effects of evaporation and seepage, which are difficult to distinguish (Asare, Palamuleni, and Ruhiiga, 2018).

Rice or other vegetation is planted or grows naturally as a forage base for crayfish culture (Figure 5.8) in some ponds. When compared to evaporation by open ponds, transpiration by plants increases water loss. Because it is difficult to distinguish between evaporation and transpiration in practice, the two processes are clubbed together.

Figure 5.8. Crayfish culture.

Source: https://www.researchgate.net/profile/Yusli-Wardiatno-2/publication/324976091/figure/fig2/AS:623174051975169@1525587680157/Net-cages-used-for-culture-of-ornamental-fish-and-crayfish-in-Kemang-Lake-Java-Photo-by_W640.jpg.

Evapotranspiration (Figure 5.9) from a reference crop is calculated using coefficients based on rice vegetative growth stages. Evaporation accounts for nearly all evapotranspiration when a crop is first planted. At complete cover, transpiration accounts for roughly 90% of evapotranspiration. Ponds should ideally be built in soil that is resistant to water movement. Water moves into and out of all ponds in practice unless they are lined with a waterproof material like plastic. Marine shrimp are a high-value crop grown largely for export in tropical developing countries. The two most frequent species of marine shrimp cultivated are Pacific white shrimp or *Litopenaeus vannamei* and black tiger shrimp or Penaeus monodon (Montes-Dominguez et al., 2021). This species has also been imported to Asia, where it is gaining popularity. Marine shrimp are usually farmed in ponds along the shore, but there is growing interest in cultivating them in interior places with saline water. Because coastal farm locations are sometimes expensive and heavily regulated, inland culture of marine shrimp is of special interest in the United States. Shrimp farming has sparked greater environmental concern than other types of pond aquaculture.

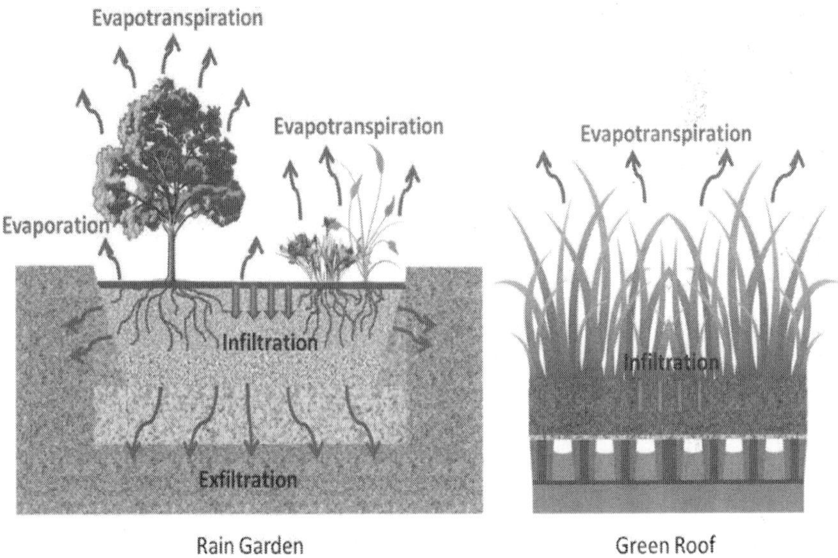

Figure 5.9. Evapotranspiration.

Source: https://ars.els-cdn.com/content/image/1-s2.0-S0048969719328293-ga1_lrg.jpg.

For all main species of marine shrimp, the basic processes for culturing are comparable. In ponds with saline water, post larval shrimp are stocked. Shrimp are raised in waters with salinities varying from 1 to 40 parts per million. The practice of acquiring broodstock from the wild and raising post larvae in hatcheries gradually superseded it. In today's hatcheries, farm-reared broodstock is frequently used. To avoid water quality degradation, nutrients from fertilizers and feeds stimulate phytoplankton development, and water exchange and mechanical aeration can be used. When shrimp infections become a problem, antibiotics may be added to the feed. Shrimp output ranges from less than 500 kg/ha in big, unfertilized ponds to more than 10,000 kg/ha in very intensive ponds with initial fertilization, significant daily meal inputs, heavy mechanical aeration, and daily water exchange. Most shrimp farms yield 1,000 to 5,000 kg per hectare per crop. Two or more crops are produced annually in tropical climates. For harvest, ponds are entirely drained, and the bottoms are allowed to dry out in between crops. To neutralize sediment acidity, agricultural limestone or lime may be applied, bottoms may be tilled to improve aeration, and sediment may be removed if accumulation is considerable. To ensure suitable salinity in inland shrimp culture, ponds are filled with salty water from wells or brine solution from seawater evaporation ponds combined with freshwater. Mineral salt additives are frequently used to address major ion imbalances in saline well water, particularly low potassium and magnesium concentrations. Inland farms typically have limited water supplies; therefore, water exchange is either not used or done through cycling of water between ponds and a reservoir. Inland pond shrimp culture follows the same methods as coastal pond shrimp culture. Shrimp culture is a minor business in the United States, and it is unlikely to grow much in the future. The high cost of coastal land, the impossibility to grow more than one crop each year in subtropical or temperate climates, and land and water usage rules are the main factors.

5.12. BRACKISH WATER

Coastal crops require constant access to brackish or seawater. To avoid salinization of freshwater, they must discharge into estuaries or the sea. By affecting local hydrology, roads, embankments, and other farm infrastructure must not alter salinity patterns. Roads at shrimp farms, for example, might block tide flow into coastal marshes, preventing them from receiving the typical flood of saline water. Coastal areas in numerous places may flood during heavy rains. Flood heights should be determined, and pond embankments should be built high enough to prevent flooding. Storm surges must also be factored into farm design to avoid pond flooding. This is especially relevant along the Gulf Coast, where hurricanes frequently create enormous storm surges. Floods, droughts, severe storms, earthquakes, and other potentially harmful natural events should be analyzed using climatological and geological records pertinent to a specific location. Seismic activity is likely in many coastal areas with shrimp farms around the Pacific and Indian Oceans. The infrastructure of several shrimp farms has been damaged by earthquakes. Tsunamis caused by earthquakes can also destroy shrimp farms, as Indonesia, Thailand, India, and Sri Lanka experienced in

December 2004. Although the tsunami of December 2004 was extremely catastrophic, it was not entirely unexpected. Tsunami data from four places with shrimp farming between 1950 and 2003 found that these events were more common in Indonesia, with the extent and water height being notably high in that country (Tanyaros and Crookall, 2011). Tsunami water heights in Indonesia and the Philippines show that any shrimp farm built near the coast, regardless of construction methods, might be wrecked by an average tsunami. The typical water height in the Caribbean region and Ecuador is substantially lower, and the chance of pond inundation can be reduced by building embankments higher than maximum projected water levels. Hurricanes will become more frequent and intense, and rainfall variability will increase, according to the Intergovernmental Panel on Climate Change (IPCC). This shows that using historical climatic data would understate the hazards of storm related damage and rainfall anomalies at a specific location. When planning projects in coastal areas, it may be practical to leave more room for error.

More than one hurricane affected Texas in the United States and the Yucatan region of Mexico every two years, whereas Belize received slightly less than one storm per two years. In the Philippines, four to eight hurricanes strike each year, with an average of roughly six. Hurricanes and severe storms are typical in these four areas, as well as many others, therefore shrimp farms should be built and equipped to withstand strong winds, exceptionally high tides, and storm surges. Rainfall in South and Central America during the rainy season fluctuates substantially from year to year due to oceanic currents known as the El Nino Southern Oscillation (ENSO) (Figure 5.10).

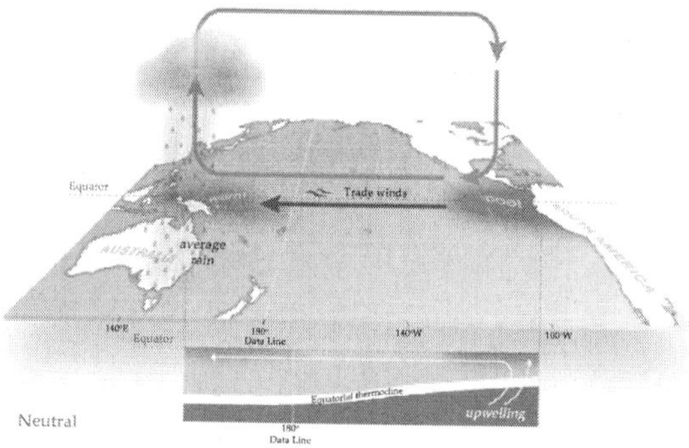

Figure 5.10. El Nino Southern oscillation.

Source: http://www.bom.gov.au/climate/enso/history/ln-2010-12/images/6-7_ENSO_12-07-12_neutral-outlines.png.

During an El Nino event, the ocean currents around the South American coast weaken, allowing warmer water from the western Pacific Ocean's tropical zone to move eastward. During El Nino years, the ocean water offshore is warmer than usual, resulting in more rainfall. El Nino occurrences are frequently followed by La Nina events, in which the sea offshore is cooler than average and rainfall is lower. There is no cycle with a defined period between incidents, despite the fact that they tend to rotate. Shrimp farmers on South and Central America's Pacific coasts should prepare for alternating heavy and light rainy seasons. The El Nino phenomenon has an impact on the United States, and farm water supplies should be chosen with the goal of avoiding salinity oscillations caused by changes in rainfall (Sazib, Mladenova, and Bolten, 2020).

When strong rainy seasons are expected, ponds should be operated to conserve salinity. In estuaries, freshwater and salt water do not mix well, and the higher salinity water sinks to the bottom. Freshwater intake can be avoided at some places by constructing pumps for deeper intake. Estuaries with heavy rainfall have more turbidity. The sediments settle quickly when water with high suspended solids concentrations is delivered to water supply canals or ponds. Excessive sediment collection in canals and ponds diminishes volume, while soft material in ponds create problems with feeding and harvesting. During years with particularly heavy rainfall, rivers flowing into estuaries and waterways inside estuaries may occasionally alter course. Pump stations for shrimp farms have had to be relocated in the past because changes in river courses expanded catchment areas, resulting in more freshwater and unacceptably low salinity. In flooded estuaries with poor water exchange with the sea, heavy rains can be helpful. Pond embankments must be high enough to prevent overtopping from high tides, storm surges, and flooding. For transport to canals and ponds, water must be elevated several meters above sea level. If the land is raised higher than it has to be, pumping costs and energy consumption will be excessive. As a result, flat areas above the highest tidal level should be preferred for shrimp farm locations. For ponds and canals, soil samples should be taken to at least the maximum depth of excavation. Delineating the construction area and taking core samples to the required depth at intersections of a 50 m or smaller grid superimposed over the site is the optimum approach. Particle size distribution, pH, and organic matter concentration should all be tested on these samples. Sulfur analysis of samples is the best way to do this. When soils with more than 0.75% sulfur are dried, they can become very acidic. Other soil parameters such as liquid, plastic, and shrinkage limits, as well as coefficient of linear expansion, should be measured by the contractor for use in designing earthworks and planning construction activities.

Submerged, suspended, or floating enclosures used to store and grow aquatic species are known as net pens and cages. In freshwater or saltwater, net pens and cages can be placed along a shore and drifting offshore.

Net pens (Figure 5.11) feature stiff frames just at the top and cages have rigid frames on all sides. Water quality in net pens is maintained by tides, surface currents, and other natural water movements, which provide a constant supply of high-quality water. Unlike flow-through systems, which sweep nutrients away from the culture unit, diluted effluents from net pens in marine systems can be returned to farm sites via tidal circulation. The quality of water surrounding and flowing through net pens has a direct impact on the health, welfare, and quality of aquatic species produced there. Importantly, producers that use net pens have limited protection for their animals from poor water quality and other environmental factors. As a result, whatever influence management techniques have on water quality or the local environment has a direct impact on the performance of cultured animals and farm economic returns. Producers are encouraged to manage activities in line with the local environment because of the close link between environmental quality, cultured animal performance, and farm income. Understanding local site features is essential for controlling agricultural activities and ensuring that a site's ability to digest and process waste solids and nutrients is not exceeded. In local contexts, net pens develop new underwater structures. Similar to designed artificial reefs, these structures and their anchoring systems provide three-dimensional habitat. Net-pen facilities can increase the total amount of available habitat as well as the complexity of that habitat.

Figure 5.11. Net pens.

Source: https://ecology.wa.gov/DOE/media/Images/WATER-SHORELINES/rich-passage-empty-may-2021.jpg?ext=.jpg.

Trawlers (Figure 5.12) are unable to access benthic habitats under and around the farms due to structures like pens and their related anchoring systems. As a result, these ecosystems are safe from the physical damage caused by destructive fishing techniques. However, these advantages are contingent on adequate net-pen farm site selection and maintenance. The evolution of net pens for fish could have paralleled the evolution of fences and other confinement technologies for terrestrial animals. Net-pen enclosures, like ponds, are said to have arisen from early attempts to cage and contain wild animals. Stone, brush, and wood were used to build these early constructions. These traps later included the use of natural and synthetic fibers, as well as netting, when new materials were invented. The rationale behind terrestrial and aquatic animal holding systems is that by enclosing live animals in the present, food can be kept for future times of scarcity. The length of time an animal may be confined in a terrestrial or aquatic enclosure is determined in part by its physiology and dietary requirements. Long periods of confinement without food will result in deterioration, disease, and death.

Figure 5.12. Trawlers.

Source: https://www.bairdmaritime.com/wp-content/uploads/1-Sharon_Anne.jpg.

Natural fibers were used to make the nets. The net bag's size and durability would have been constrained as a result. Although net-pen materials and designs have advanced quickly since WWII, locally available and low-cost materials remain critical. Many net pens were made of wood sandwiched around some type of flotation substance as recently as the 1980s. In Asia, bamboo is still a popular net-pen material for floating collars. Following WWII, various technological advancements made larger net pens possible. First, the advent of new synthetic fibers like nylon, as well as enormous mechanized looms, allowed for the production of very large, durable net panels. Second, new materials like galvanized steel and high-density polyethylene allowed for significantly larger floating collars to be built. The concept of production intensity remains valuable, even if the units of measurement used to quantify intensity and the relative classification of low, medium, and high-intensity processes may need some alteration when applied to net pens. When comparing different output levels or farm management practices on the same site, this is especially true. The level of environmental effect or risk connected with a net-pen facility can be directly influenced by the intensity at which it is operated. This is predicated on a direct relationship between waste loading and production intensity in the surrounding ecosystem, and at some point, the ecosystem's ability to cope with that loading is exceeded. Because there is a direct link between the surrounding environment and the performance of the cultured fish, more environmental risk translates into greater business risk in the case of net pens.

Environmentalists frequently stated these issues as reasons why net-pen aquaculture should not be permitted. Many farms failed due to pollution and poor fish performance due to the combination of low-energy areas and the early stage of development of net-pen farming methods. Temporary enclosures directly immersed in aquatic ecosystems are known as net pens. One of the most effective farm management methods for minimizing environmental problems is site selection. Significant technological advancements in net-pen design and materials have allowed systems to be placed in progressively high-energy and exposed locations. In open-ocean settings, a variety of offshore net-pen designs have been devised and tested. These designs employ new technology and materials to ensure structural integrity during extreme weather events like hurricanes and typhoons. Although improved net-pen designs and materials have demonstrated their ability to withstand harsh circumstances, open-ocean operations have yet to be proven economically. The benefits of these technologies' development, on the other hand, have been recognized in near-shore situations, culminating in a general industry-wide trend toward larger facilities in deeper water. Because all net-pen facilities employ nets to contain fish, they appear to be the same on the surface. However, there are a surprising number of species cultivated in net pens around the world. Salmonids, particularly Atlantic salmon, have received the majority of public attention in net-pen aquaculture (Love, Fry, Cabello, Good, and Lunestad, 2020). Rainbow, trout, sea trout, coho salmon, and Chinook salmon are among the salmonids raised in net pens.

Some of these species' aquaculture is still reliant on wild-caught juvenile fish. For example, yellowtail mackerel and Japanese jack mackerel have been harvested and cultured in net pens in Japanese coastal waters for more than 60 years, and more recently in Korea and Spain; several species of tuna (Thunnus spp.) in Australia, Canada, Croatia, Japan, Malta, Mexico, and Spain; and several species in Southeast Asian countries. Water exchange is required in flow-through aquaculture systems to maintain adequate water quality for fish production and to collect and remove metabolic wastes. Water for flow-through facilities is typically diverted from streams, springs, or artesian wells and gravity-fed through the farm. Pumped water from wells or other sources is costly and is only utilized in small hatcheries. Various public authorities control water diverted from springs or surface sources for flow-through aquaculture, based on each state's individual water regulations. Surface water diversion is considered a non-consumptive use, while in some areas, pumping groundwater from a well is deemed a consumptive use. The discharge of a high-volume, dilute effluent from flow-through aquaculture plants makes it difficult to estimate environmental impacts and severely limits treatment options.

5.13. COMMERCIAL PRODUCTION

Commercial production of rainbow trout or Oncorhynchus mykiss in concrete raceway systems is sometimes associated with fish production in flow-through culture systems in the United States (Welker, Overturf, and Abernathy, 2019). Brook trout or Salvelinus fontinalis and brown trout or *Salvelinus niger* are two more cold water fish species that are generated in flow-through systems. Public hatcheries produce additional trout species in these systems, but the amount is negligible compared to commercial rainbow trout production. Freshwater stages of salmon are produced using flow-through systems. On a small scale, flow-through systems are also utilized to produce warmwater fish like catfish or *Ictalurus* spp. and tilapia.

Cool water species like yellow perch (Figure 5.13), hybrid striped bass, and various species of sturgeon have recently been produced using flow-through systems. The most prevalent aquaculture production techniques for rainbow trout and other salmonids in the United States are flow-through systems. Linear earthen and concrete raceways, as well as tanks made of various materials, are examples of flow-through systems. The most prevalent are concrete raceways, but circular rearing tanks are also employed in flow-through systems, mostly for broodstock production. A raceway is essentially a water-carrying flume. A tank or a group of rectangular tanks with water flow along the long axis make up a basic raceway producing system. Water flow will resemble plug flow in an ideal raceway, with uniform water velocity across the tank cross section. Water velocities will vary over the breadth and depth of the raceway due to friction losses at the tank water and air-water boundary layers. Water velocities are highest at mid-depth, with slightly lower speeds at the air-water interface and significantly lower velocities along the raceway bottom. More fish can be developed in first-use water if raceways are built in parallel rather than in series, but the intensity of production per unit of water flow will be reduced unless oxygenation or aeration is used. In fact, most flow-through aquaculture facilities employ a combination of parallel and series raceways. The design of a flow-through aquaculture production facility is influenced by available water supply and quality, space availability, and geography. Groundwater and surface waters that are gravity-fed to the farms serve as water sources for flow-through aquaculture.

Figure 5.13. Yellow perch.

Source: https://www.floridamuseum.ufl.edu/wp-content/uploads/sites/66/2020/02/Plate-178.jpg.

CHAPTER 6

SUSTAINABLE FISH FARMING

CONTENTS

6.1. Introduction 122

6.2. Oyster 122

6.3. Water Quality 123

6.4. Mechanism 125

6.5. Food Production 126

6.6. Microbes 132

6.7. Sustainability 134

6.8. Nutrition 135

6.9. Microalgae 136

6.10. Natural Resources 138

6.11. Diseases 143

6.12. Ammonia 144

6.1. INTRODUCTION

Seafood, being a popular high-protein diet, includes omega-3 fatty acids, which are essential for human growth. Seafood is high in vitamins and minerals such as vitamins A and D, phosphorus, magnesium, selenium, and iodine, while being low in calories, total fat, and saturated fat. Fish has been found to provide a variety of health benefits. Seafood is a low-fat, nutritious alternative to beef, chicken, and pig, with significantly greater omega-3 fatty acid levels than vegetarian diets. Fish oil, in particular, contains omega-3 fatty acids, which are crucial for infants and babies to build the brain. The most important drivers of demand for high quality and healthy seafood products are population increase and economic development trends. Aquaculture has provided a bridge between expanding demand and conventional seafood sources, as wild capture fisheries have reached their maximum sustainable harvest capacity. In today's worldwide marketplace, farmed seafood accounts for around half of total production. However, the United States has a seafood deficit, importing more seafood than it can produce to fulfill consumer demand.

6.2. OYSTER

Natural oyster reefs (Figure 6.1) provide many of the same ecosystem functions as oyster shellfish aquaculture. Rearing shellfish in high numbers in shallow water with rich phytoplankton concentrations, unlike other finfish farming approaches, can have good environmental consequences and may enhance biodiversity. As the world seeks to feed a population of over 7 billion people, the technique of aquaculture has come under increased scrutiny. This criticism is leveled regardless of whether aquaculture is viewed as a source of revenue for developing countries or a potential food source. Aquaculture is the world's fastest expanding food production business, but its long-term viability is uncertain. It was mentioned that a delicate balance between food security and production costs that are environmentally friendly must be reached. Fishermen and seafood processors are losing their jobs as our most vital fisheries collapse. Aquaculture, particularly marine aquaculture, holds the key to solving this problem. Although, technical advancements have made it possible to cultivate aquatic species in a safe, economic, and environmentally sustainable manner, sustainability appears to be the key to long-term financial and environmentally sound aquaculture methods. Given that the majority of fish we eat are farm-raised, with over 100 species farmed worldwide, many culturing techniques have been adopted, ranging from traditional clay ponds to high-tech tank systems, and each culture method/technique has its unique environmental footprint (Kumar and Sharma, 2020). Although most aquaculture facilities strive for stress reduction, good health, and rapid growth, many larger intensive aquaculture systems raise stock in stressful environments with minimal ecological balance. Managers risk increased production costs, animal deaths, and habitat deterioration in areas where liquid waste is discharged.

Figure 6.1. Natural oyster reefs.

Source: https://upload.wikimedia.org/wikipedia/commons/thumb/e/ec/Oyster_reef_Hunting_Island_SC.jpg/800px-Oyster_reef_Hunting_Island_SC.jpg.

6.3. WATER QUALITY

Eutrophication, benthic enrichment, habitat modification, erosion, disease, water quality, and successful application of best management practices (BMPs) are some of the other areas of concern in aquaculture. Bacteria and viruses become significantly more relevant in respect to aquaculture water quality concerns as nutrients and phytoplankton are added. Bacterial diversity and nutritional content have a direct relationship. In nutrient-rich waters and sediment surface layers, naturally occurring bacteria from the environment and the intestines of cultivated fish populations thrive. Humans can be infected by zoonotic viruses through a variety of routes, including food and contact with contaminated environments. In non-native stocks, where introduced species and hybrids may introduce novel viral strains, viruses are a particular problem. Even if the risk of viruses being introduced is lowered, virus outbreaks are not uncommon. The spread of Vibrio bacteria in receiving water bodies appears to have been aided by uncontrolled growth inside agriculture activities. Many of aquaculture's difficulties could be solved with integrated aquaculture. Because no species exists in a vacuum, stocking production facilities with complementary species is a sensible strategy to integrate diverse species while also enhancing production for a given area. For example, grass carp or other herbivores could be maintained alongside primary stock to limit algae and plant growth. Suspension feeding bivalves are beneficial species for phytoplankton filtering. It was discovered that passing water over 11 oyster rafts reduced phytoplankton counts by 94%. Secondary stocks are not only useful for managing water quality, but they are frequently valuable food crops as well (Jones, Preston, and Dennison, 2002). Integrated multi trophic aquaculture systems not only increase profit and reduce costs, but they also improve economic stability and give more acceptable management approaches.

Some studies showed how to run an aquaculture enterprise efficiently while minimizing environmental impact and maintaining a profitable output. BMPs are the most technically and economically practical strategies for reducing environmental impacts and lowering operating costs for aquaculture facilities. One of the main goals is to talk about effluent treatment systems that reduce organic matter (OM), suspended particles, and nutrients in order to keep receiving waterways clean. Selecting a site with acceptable soils and an adequate water supply, as well as maintaining reasonable organism densities and feeding rates, are the best ways to prevent soil and water quality problems. Liming, fertilization, and aeration are examples of secondary management practices for preventing soil and water quality imbalances. It's critical to combine older, proven ways with new and inventive ideas when it comes to implementing aquaculture BMPs. Aeration, settleable solid removal, pH changes, and anaerobic denitrification are some of the other management strategies used. Even if this type of system isn't as profitable as producers would want, the basic concepts can easily be extended to a variety of aquaculture systems. Unfortunately, unless discharge laws are tightened or the systems are demonstrated to be economical, there is little incentive to create them.

6.4. MECHANISM

Several mechanisms that are typical in undisturbed habitats keep the extremely changeable nature of any aquatic ecosystem in a delicate balance. This delicate balance can be upset when anthropogenic stressors are introduced into the environment. The consequences of seepage and discharge off farms can disrupt the healthy state of aquatic ecosystems within entire watersheds as a result of increased aquaculture activity and related farming practices. Elevated nutrient concentrations might have negative consequences, especially in receiving water bodies with limited water exchange, such as inlets. Changes in microbial growth and community composition are responsible for the negative impacts. Poisonous environments are frequently caused by damaging algae blooms, de-oxygenation of water and sediments caused by excessive microbial development, and the transfer or concentration of toxic substances through the food web. Rivers, sewage, agriculture, and fish farms all contribute to the growth of dissolved and particulate pollutants in estuaries and coastal environments. Because these dissolved and particulate components are high in carbon, nitrogen, and phosphorus, they supply nutrition for phytoplankton and bacteria.

The carp or *Cyprinus carpio* and mullet or *Mugil cephalus* (Figure 6.2) were the earliest culture species. For ages, this activity has fueled human demand for fish products, and it is now a major global industry (Papoutsoglou, Petropoulos, and Barbieri, 1992). Over time, this approach has become increasingly technical with the goal of making work easier while also increasing production. This activity has now evolved into a full industry that deals with both supplements and the actual product. Aquaculture now accounts for one-third of all seafood consumed on the planet. With this tremendous rise in global production, the aquaculture business is now one of the fastest expanding food production sectors. However, due to a large use of water, land, and feed, the aquaculture industry's rapid rise has caused environmental damage. Furthermore, these requirements are linked to other consequences such as pollution, soil salinization, nutrient loading, natural habitat loss, overexploitation of groundwater supplies, and disease introduction and transmission. Increased production in aquaculture necessitates feeds, energy for the cultured species, and almost always comes from catch, putting the fastest-growing sector in food production in jeopardy of continuing to provide increasing yields in a sustainable manner.

Figure 6.2. Mugil cephalus.

Source: https://www.fishbase.in/images/species/Mucep_u4.jpg.

6.5. FOOD PRODUCTION

Current aquaculture food production systems are clearly beneficial from a commercial standpoint, but environmental considerations are always overlooked. The relationship between aquaculture and the environment is complicated, particularly when it comes to biodiversity. There have been several examples of both positive and negative consequences, but there has yet to be a solution that allows for the creation of a relationship between food production and the environment. This approach must be founded on ecological ideas such as extreme resource efficiency and the nutrient and waste cycles, as well as resource-use optimization. The efforts and perceptions of the environment vary by country, yet the problem must be addressed holistically in the end. Fish studies, on the other hand, tend to focus on species with current commercial value, leaving species with lower market value to be overlooked. Several freshwater native species found in Central and South America fall under this category. Some attempts to cultivate native species have happened mostly in rural regions, where, in addition to enhancing species conservation, protein contributes to the community's food.

Growing white fish or *Chirostoma estor* (Figure 6.3) with the goal of repopulating some places where their introduction has declined is one of the most interesting case studies in Mexico. The native Central American cichlid or *Cichlasoma urophthalmus* (Figure 6.4) is the most well-documented, with numerous studies supporting its culture. *Herinchthys cyanogutattus* (Figure 6.5), on the other hand, was discovered in the Pánuco river basin. After becoming acclimated, the stock was used to conduct density tests of individuals for culture and ideal thermal for its production. *Procambarus digueti* has serious ecological issues because it has been harvested for food and medicinal purposes since pre-Hispanic times. The alteration of their habitat has added to the intense environmental pressure. The goal of many studies was to find the best growing conditions for *P. digueti* growth in intensive production systems in terms of load capacity and nutrition. The managed production of this species will lessen fishing pressure and will be able to help repopulate areas where the natural population has declined. If aquaculture potential investigations are conducted using an ecophysiological and bioenergetics approach, prediction models of population development under various environmental conditions can be produced.

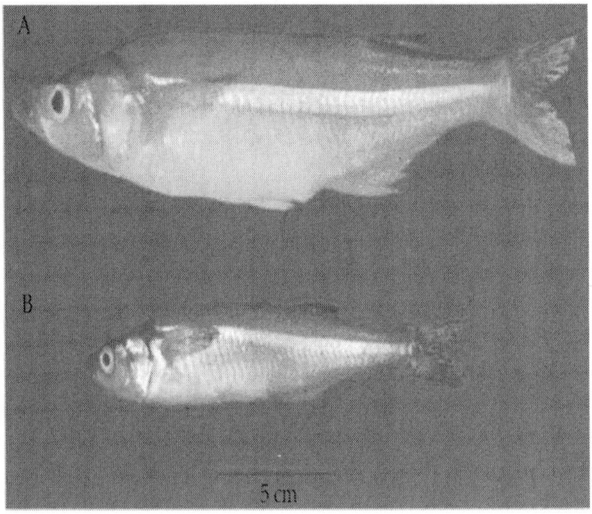

Figure 6.3. Chirostoma estor.

Source: https://static.inaturalist.org/photos/9199005/medium.png.

Figure 6.4. Cichlasoma urophthalmus.

Source: https://d9-wret.s3.us-west-2.amazonaws.com/assets/palladium/production/s3fs-public/thumbnails/image/mayan%20cichlid.png.

Figure 6.5. Herinchthys cyanogutattus.

Source: http://www.fishtanklab.com/pictures/257/500.png.

Studies on the biological characteristics of fish often focus on species that are currently commercially valuable, ignoring species that are not.

This is a diverse and small cyprinodontoid fish family that is restricted to Mexico's central plateau, with its dispersion center in the well-isolated Lerma basin. In Querétaro, four Godeidae species have been identified: three are extensively spread in the Lerma basin. They are *Goodea atripinis* (Figure 6.6), *Xenotoca variata* (Figure 6.7), and *Goodea gracilis* (Figure 6.8), and *Girardinichthys multiradiatus* (Figure 6.9) is restricted to a single body of water in the municipality of Amealco. Sexual dimorphism, strange courtship rituals, and viviparity, taxonomic features, ethology, biology, and trophic ecology are among the topics studied by scientists (Garcia-Trejo et al., 2015). Aquaculture is a key food-producing business that is steadily expanding in tandem with rising population density and other sectors' land-use requirements. Aquaculture must transition to intensive or semi-intensive operations, effective and sustainable resource management, and sustainable environmental stewardship in order to retain growth. This frequently necessitates the use of technology that improves production efficiency while avoiding rivalry for space and resources with other operations like agriculture and ranching. Aquaculture operations must be environmentally sustainable and minimally harmful, meet quality and safety requirements, and allow for efficient use of space and natural resources as well as expansion opportunities. Current procedures must embrace technology alternatives that reduce environmental effects and are efficient without harming the health and growth of stock organisms. Using biofloc technology is one alternative. Biofloc is made up of nitrifying bacteria, OM, inorganic flocculants, and suspended algae that naturally grow in pond water. These components feed the growing stock and encourage the direct usage of nitrogenous substances found in feces, urine, and food waste. With the addition of carbon sources and continual aeration, nitrifying bacteria activity increases, maintaining or dramatically improving water quality during cultivation. As a result, the vast amount of water needed for intensive aquaculture is considerably reduced. The use of biofloc in the cultivation of the Malaysian river prawn *Macrobrachium rosenbergii* (Figure 6.10) is an example (Pérez-Fuentes, Pérez-Rostro, and Hernández, 2013).

Figure 6.6. Goodea atripinis.

Source: https://www.biolib.cz/IMG/GAL/326256.jpg.

Figure 6.7. Xenotoca variata.

Source: https://www.fishbase.se/images/species/Xevar_u0.jpg.

Figure 6.8. Goodea gracilis.

Source: http://www.borstein.info/webpics/Photocredit/TedJudy/goodea_gracilis.jpg.

Figure 6.9. Girardinichthys multiradiatus.

Source: http://www.goodeidworkinggroup.com/sites/default/files/goodeiden/IMG_3459.JPG.

Figure 6.10. Macrobrachium rosenbergii.

Source: https://upload.wikimedia.org/wikipedia/commons/thumb/a/a4/Macrobrachium_rosenbergii.jpg/1024px-Macrobrachium_rosenbergii.jpg.

6.6. MICROBES

The microorganisms that live in biofloc systems (Figure 6.11) are usually found in natural aquatic environments. Their presence is directly proportional to two environmental variables: the amount of solar energy absorbed by the system and the amount of OM and carbon sources that enter the system. Colonies of bacteria in a biofloc system rely on OM in the system for survival and multiplication, and to a lesser extent, the intensity of sunlight. Bacteria are also reliant on the system to provide regular aeration since they use enormous amounts of oxygen, which is directly linked to the bacteria absorbing carbon from the system.

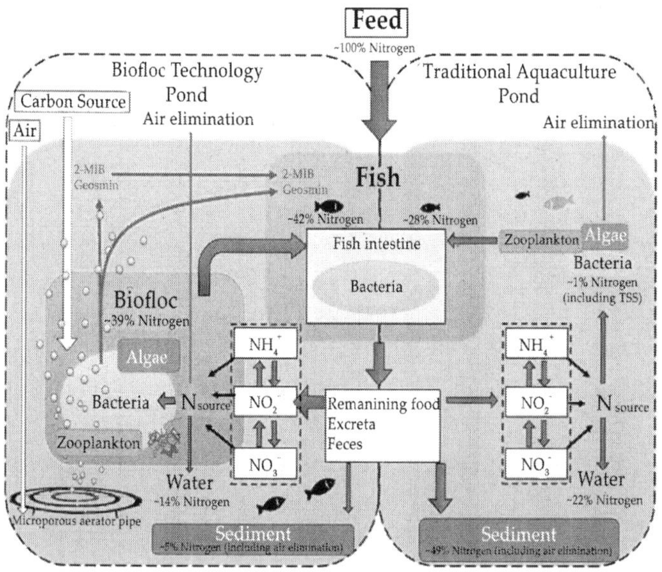

Figure 6.11. Biofloc systems.

Source: https://www.exeley.com/exeley/journals/polish_journal_of_microbiology/69/4/images/10.33073_pjm-2020-049-f001.jpg.

Carbon is a source of energy for expansion and reproduction. The carbon concentration in the system must be kept at a C/N ratio that allows bacteria to reproduce with inorganic nitrogen at a maximum concentration of 200 mg/L (Avnimelech, 1999). This value indicates that the nitrogen control system is working efficiently and effectively. High nitrogen concentrations will usually disrupt the system's balance and harm stock species, particularly shrimp. Nitrogen compounds are eliminated in a biofloc system based on bacteria and algae because bacteria boosted ammonium absorption and better controlled waste products. This does not rely on the amount of sunshine to function properly. In a microalgal biofloc system, production is directly dependent on sunshine; excessive lighting might result in an overabundance of algae, resulting in low oxygen levels at noon. As a result, a bioremediation system based on microalgae and bacteria will be more effective because it will be feasible to maintain an efficient balance between nitrifying bacteria and algae to maintain a proper level of nitrogen, a balanced C/N ratio, and sunlight.

The microorganisms that make up the biofloc and break down the nitrogen compounds that pollute fish pond water require energy to function. Depending on the type of the organisms present in the biofloc system, there are three possible energy sources in aquatic biofloc systems. The most essential source of energy for phototrophic microorganisms like algae and vascular plants is sunshine. The second source of energy is inorganic chemicals, which are used by microbes to obtain energy by oxidizing reduced versions of simple molecules, particularly nitrogen. Nitrogen is converted to nitrite and nitrate in fish aquaculture by metabolizing organic nitrogen and ammonia. Organic substances, which are altered by microorganisms that receive energy from the metabolic oxidation of organic carbon and convert it to carbon dioxide, are the third source of energy. Because of the abundance of energy sources available, both chemotrophic and phototrophic microbes naturally consume and reduce nitrogen concentrations in the water. This is also an essential function of the microorganisms. Proteins are synthesized from nitrogen sources using transformed energy. The continual use of oxygen during the night is caused by an excess of primary production. Productivity will suffer on cloudy days, lowering water quality. Because water physicochemical factors affect yield per unit volume, efficient operations with biofloc aquaculture systems rely on keeping them within the tolerance range of grown stock. They are commercial species stock, microalgae that interact and function as bio filtrates with oxygen demand and, like commercial stock, produce metabolites, and bacteria that transform nitrogenous metabolites used by planktonic microalgae. Water quality in aquatic systems is influenced by a variety of biological and chemical activities that occur in the aquatic environment.

6.7. SUSTAINABILITY

Sustainability in the aquaculture business is determined by a number of elements, including the lowest possible production cost in relation to yield, unanticipated environmental conditions that affect the farm, and farm management techniques. These elements are intertwined and have a synergistic influence on the issue of sustainability. For sustainable farming, live feeds must be prioritized as the most basic requirement for larval rearing and fry production. Having to rely on imported sources of live or inert feed will raise production costs. As a result, it is economically vital to continue screening, stocking, and preserving some indigenous species as a viable choice for live feed production. Because newly hatched fish and shrimp larvae still have an underdeveloped digestive system and lack enzymes, live feeds are an important part of their basic diet. They are still too immature to produce or convert their own essential nutrients from any pre-cursor gained through eating. They require a ready-to-eat diet that has immediately available nutrients that can be absorbed by their digestive system. Many species have been proposed or tested for their potential as live feed. All of the test animals were largely zooplankton in nature and had to eat live food. Live feeds must be available at all times as the initial diet for larval rearing and fry generation.

6.8. NUTRITION

In terms of nutritional value, traditional live feed such as brine shrimp and rotifers are inappropriate as live feed when compared to copepods. Because polyunsaturated fatty acids (PUFAs) are insufficient in *Artemia* sp., it must be fortified before feeding. Rotifer, on the other hand, has a low nutritional value and is small in size. Copepod diets, on the other hand, have been shown to increase larval marine fish growth when compared to rotifers *Brachionus plicatilis* or Artemia (Dhont et al., 2013). Using a paracalanid as an example, the possible use of copepods as live feed has been emphasized due to their superior fatty acid content. They boosted the cultured organism's quality, especially the larval stages. Copepods were found to be superior to other live feeds such as brine shrimp and rotifers. When compared to typical live feeds, they have the right ratio of docosahexaenoic acid (DHA) to eicosapentaenoic acid (EPA), which improves the growth and survival of fish larvae. Nonetheless, due to the high expense of operation, copepods should be utilized as a supplement rather than a primary feed in larval rearing unless they are used for particular high-valued commercial reef fish. The nutritional quality of harpacticoid copepods as live feed has been established, and the idea of employing copepods, particularly harpacticoids, as an alternative live feed in fish larval rearing has been underlined. When compared to Artemia, *Macrosetella gracilis*, a planktonic harpacticoid copepod, has a greater food quality. Despite these encouraging discoveries, commercial hatcheries continue to employ rotifers and Artemia as live feeds since copepods are not currently cultivated at high enough densities to be commercially viable. In terms of culture conditions, it was discovered that a salinity of 35 PSU and a temperature of 25°C were optimal for high production of a tropical harpacticoid copepod, *Pararobertsonia* sp. (Cutts, 2003). If kept in small containers, fluctuations in salinity, pH, and temperature in the culture vessel would have a significant impact on copepod reproduction and population increase. Nonetheless, adopting a tray-culture approach to manufacture harpacticoid in large quantities for hatchery use has been proposed.

When utilized as live feed, another copepod group, the Cyclopoida, *Apocyclops dengizicus* (Figure 6.12), was found to promote the survival and growth of *Panaeus monodon* post larvae. Cyclopoids are omnivorous and can be given a variety of feeds, including phytoplankton alone or in combination with yeast or other foods. There is still a scarcity of information on the utilization of copepods in aquaculture, notably from the tropical *Apocyclops sp*. In Malaysia, for example, *A. dengizicus* has been recorded in abundance and could be used as a live feed for shrimp post-larvae.

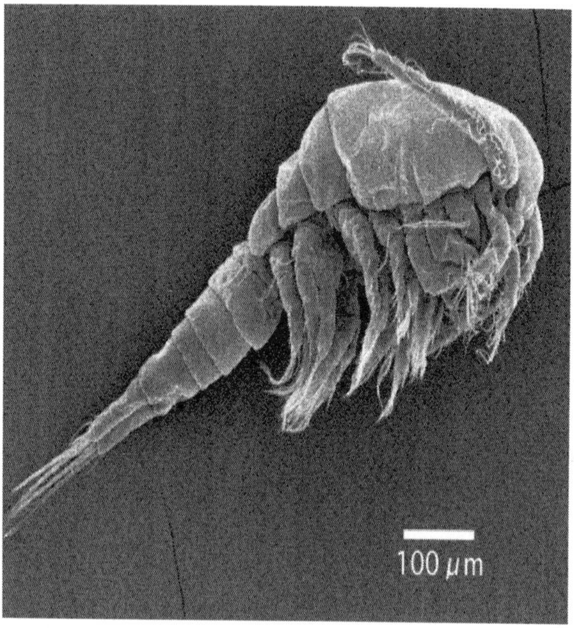

Figure 6.12. Apocyclops dengizicus.

Source: https://invasions.si.edu/SpImages/Apocylops_dengizicus.bmp.

6.9. MICROALGAE

Microalgae (Figure 6.13) are a diversified collection of unicellular autotrophs that can be found in practically any body of water. Microalgae include a variety of unique chemicals, and their nutritional importance in aquaculture has been emphasized. Microalgae production is required in the hatchery because it provides a basic and nutritious diet for live feed, particularly zooplankton. However, because of the large personnel, space requirements, and operation costs, which are frequently tied to the cost of the energy consumed, mass production is often expensive. A good technique for managing the culture environment, especially during the microalgae production process, would reduce operational costs. Through photosynthesis, light is essential for the development of microalgae. It is one of the key environmental elements that influences the phototrophic growth and productivity of microalgae. Natural light or fluorescent tubes can be used to provide the most effective radiation that the pigments of the microalgae can absorb. To penetrate through the culture at a bigger volume, light intensity must be raised. Extreme light intensity, on the other hand, can cause photo-inhibition, which lowers photosynthetic rates and growth. Furthermore, in microalgal culture, warming caused by artificial or natural illumination should be avoided. The most commonly used light intensity is 1,000 lux, which is appropriate for Erlenmeyer flasks, but larger volumes of microalgal cultivation require 5,000–10,000 lux. In the light/dark (LD) cycle, the duration of illumination can be changed to enhance or increase photosynthesis of microalgae (Perumal et al., 2012).

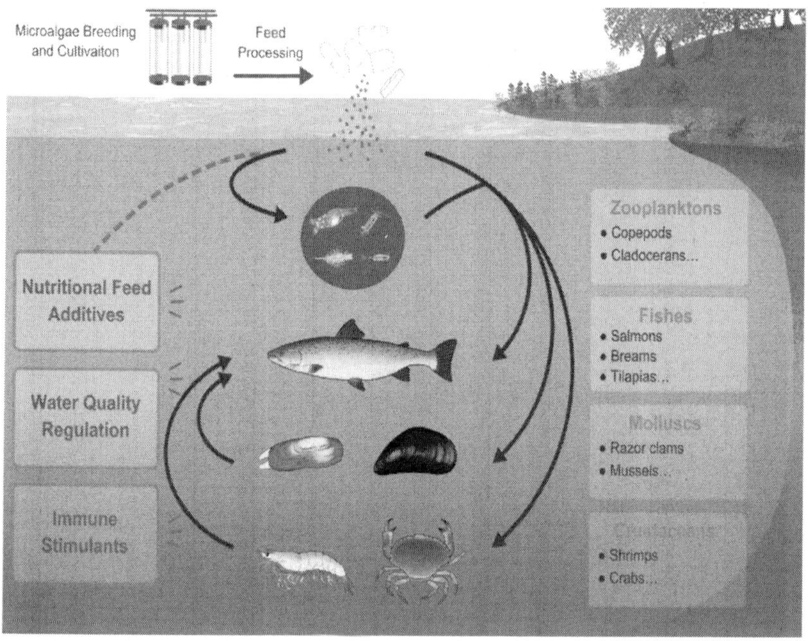

Figure 6.13. Microalgae.

Source: https://www.frontiersin.org/files/Articles/590431/fbioe-08-590431-HTML/image_m/fbioe-08-590431-g001.jpg.

For some microalgae, photosynthesis rates can be boosted exponentially by increasing LD frequencies, with a longer dark period in respect to the light period improving photosynthetic efficiencies. The use of nutrients in the culture vessel is also affected by the illuminations. The development of moist-microalgae concentrates is a cost-effective and nutritionally appropriate alternative to the costly upkeep of live microalgae. It is thought to make hatchery procedures easier, and it has shown promise in the aquaculture business. Microalgae concentrates can be kept moist and stored at a low temperature (LT) to maintain their high nutritional content and cell viability. The growth rate of juvenile pacific oysters or *Crassostrea gigas* fed different algal pastes was much higher than that of oysters fed other diets. The survival rates of tiger prawn or *Penaeus monodon* larvae fed with live microalgae were found to be identical when concentrates of *Chaetoceros muelleri* and *Tetraselmis pseudonana* were refrigerated for 6 weeks at 4°C (D'Souza et al., 2000). Supplementing bivalves, oysters, and scallops with microalgae concentrates resulted in growth rates comparable to live microalgae. The majority of the demand for mariculture feed in Japan is met with live, fresh microalgae that is thickly concentrated and easily stored at 2–4°C for 1–8 weeks with a long shelf life.

6.10. NATURAL RESOURCES

Natural resources are extensively exploited for aquaculture purposes at a high level. The intended location for brackish-water aquaculture ponds is always coastal land and mangrove forests. This area's water source, which is always from a local river estuary and lagoon, is also used as a source of live feeds. However, supply is always seasonal and may become scarce at any time due to a variety of reasons and natural events. Keeping native species is seen to be more cost-effective and practicable. It will be demonstrated that microalgae can be used as an enrichment element for live feed copepods. Unpredictable mortalities, which may be caused by unfavorable interactions between fish and harmful microbes, impeded productivity. Plant-based formulations are the least expensive of the alternatives, and many of them have an appropriate protein profile and long-term availability. Oilseeds, particularly soybean and grain products, have a lot of potential as fish feed alternatives. Soybeans are high in protein and are the most widely consumed plant protein source worldwide. SBM or soybean meal is already a popular protein component in fish feed. Inclusion of various vegetable proteins, such as SBM, in the diets of fish at levels greater than 20% might produce intestinal problems such as pathomorphological alterations in the distal intestinal epithelium and diarrhea, which are sometimes induced by antinutritional substances found in SBM (Bonvini et al., 2018). In first-feeding rainbow trout, adding probiotics to starter meals tended to promote SBM utilization. In this perspective, two of the most significant issues in fish aquaculture facilities are disease control, particularly during the initial stages of life, and nutrition improvement through optimal food usage, particularly for novel fish species.

Even though yeast represents less than 1% of the total microbial isolates in the host, it can provide a significant physiological contribution beyond that of probiotic bacteria; in fact, yeast cell sizes can be a 100-fold bigger than those of bacteria. Unlike bacteria, yeast cells use a vast range of chemical molecules, both simple and complex. This is due to yeast's broad metabolic potential, which is reflected in the production of several enzymes. Polyamines produced by yeasts are also involved in the development of fish larvae's digestive tracts. Additionally, certain yeast species and their components, such as beta-glucans and manno-proteins, might boost the host's immunological and antioxidant systems. Understanding the role of yeast microbiota in fish health and nutrition may help to improve sanitary conditions as well as fish productivity.

Yeasts can be found in a variety of natural habitats, including soil, freshwater, and ocean. The concentrations and types of organic materials available determine their quantities and species distributions. Marine yeasts are involved in a variety of ecological processes in the sea, particularly in estuarine and nearshore environments, including the decomposition of plant substrates, nutrient recycling, and the biodegradation of oil/recalcitrant compounds, as well as being a component of the microbiota of marine and aquaculture animals. The fact that yeasts have exceptional metabolic potential contributes to their functional diversity. This potential is exploitable, but the great majority of it has yet to be uncovered. As reagents, cell proteins, vitamins, pigments, and enzymes, certain yeast compounds have important biological utility. Several yeast species have been discovered in various marine environments. *Debaryomyces hansenii* and *Candida spp.* are common ascomycete yeasts that can be found in oceanic and other aquatic habitats. In oligotrophic oceanic water, basidiomycete yeasts often make up the bulk of the entire yeast population. *Cryptococcus*, *Rhodotorula*, and *Sporobolomyces* are *Basidiomycete* yeasts that are widely distributed over oceanic regions.

Debaryomyces hansenii (Figure 6.14) is a non-pathogenic, halotolerant yeast that can grow in a range of settings, including the marine environment and the intestines of fish. *D. hansenii* has been identified as one of the most commonly isolated yeasts linked to fish. This species thrives in seawater, which could explain why it's so common in fish guts. *D. hansenii* has a significant biotechnological advantage over *Saccharomyces cerevisiae* in that it is osmotolerant and can grow in conditions containing up to 4 M NaCl, whereas *S. cerevisiae* can only thrive in media containing < 1.7 M NaCl (Stratford, Steels, Novodvorska, Archer, and Avery, 2019). Because of its great enzymatic potential, *D. hansenii* has been widely researched. Several enzymes of biological and industrial significance, including inulinase, protease, superoxide dismutase (SOD), lipase, catalase, and galactosidase, have been discovered and described in this yeast. Phytase from *D. castellii* and glucosidase from *D. pseudopolymorphus* have also been found. The capacity of *D. hansenii* to manufacture galactosidase has proved effective in the reduction of raffinose oligosaccharides, which are anti-nutritional factors for humans and fish. In animal models, *D. hansenii* SOD has been recommended as a therapeutic anti-inflammatory drug. *D. hansenii* is one of the yeast species chosen for full sequencing because of these traits.

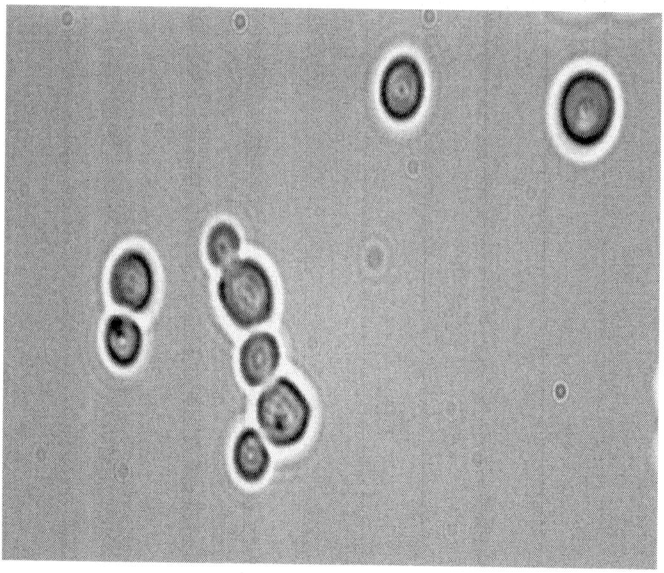

Figure 6.14. Debaryomyces hansenii.

Source: https://wineserver.ucdavis.edu/sites/g/files/dgvnsk2676/files/inline-images/2768e%20cs5.jpg.

It is well acknowledged that every ecosystem contains both cultivable and viable species that cannot be cultivated in the laboratory. Only about 1% of the microbial species found in the ocean can be grown. Several molecular approaches for studying natural materials have been developed in recent decades. Amplification of segments coding for the 16S rRNA gene has emerged as a powerful method for investigating bacterial diversity. In molecular microbial ecology, denaturing, or temperature gradient gel electrophoresis (PCR-DGGE/TTGE) techniques have been used to investigate the genetic diversity of bacterial communities found in the fish gut. Using RNA extracted directly from the samples, these approaches were also used to define the major active bacteria in the intestine of distinct rainbow trout groups. However, one significant drawback of PCR-based approaches is their limited sensitivity, which can only identify about 1% of all species. The use of rRNA gene fingerprinting, on the other hand, necessitates sequencing of the cloned bands in order to precisely identify the community's members. Because amplicons from different species may migrate to the same places, or because numerous gene copies with intra-gene variations exist in one species, sequencing is required. The use of molecular techniques to the study of yeast communities has been limited to the research of food matrices. Yeast communities have been examined using PCR-DGGE and amplification of a part of the yeast 26S rRNA gene, and phylogenetic relationships of Saccharomyces species have been established using PCR-TTGE (Ubeda-Iranzo, Fernández, and Briones, 2008).

High-throughput sequencing approaches, such as pyrosequencing (Figure 6.15), have recently been demonstrated to be quick and effective ways for identifying individuals of complex populations. In general, two methodologies can be used: diversity studies based on ribosomal gene (rRNA gene) amplicon sequencing and metagenomic studies including shotgun sequencing of whole-community DNA. While sequencing ribosomal amplicons is less expensive since only one gene is sequenced, the metagenomic technique sequences all DNA genes, exposing the microbial community's roles. These high-throughput sequencing methods produce thousands of sequences per sample in a matter of days, which must be examined with bioinformatics tools. The presence of various yeast species in the gastrointestinal tract of healthy fish has been well documented, and yeast has been proven to be a significant component of the fish gut microbiota. The halotolerant yeast *D. hansenii*, in addition to *S. cerevisiae*, has been deemed a promising probiotic candidate in fish aquaculture. Because the number of reports of *D. hansenii* use is growing, we can learn more about the yeast's ability to boost fish larvae and juveniles' growth, survival, and gut maturation, as well as their immune and antioxidant systems. The fact that a modest yeast population can generate favorable effects in the host may be due to the fact that yeast cells are 100 times larger than bacterial cells. Probiotic benefits vary widely depending on the microbial species, isolation source, experimental concentration, and, finally, the fish rearing environment. However, because immune system stimulation or immunomodulation are regarded as crucial processes sustaining probiosis, one of the most commonly observed adverse effects in the host is an enhancement in immunological response. Because they contain components including glucan, mannoproteins, chitin, and nucleic acids, yeast has immunostimulatory qualities. Recent research has demonstrated that dietary *Saccharomyces cerevisiae* has a positive effect on fish. In *Uronema marinum*-infected olive flounder or *Paralichthys olivaceus*, yeast supplemented diets increase growth, feed efficiency, blood biochemistry, survival rate, and non-specific immune responses. In juvenile rainbow trout challenged with *Yersinia ruckeri*, a meal supplemented with *S. cerevisiae* treated with beta-mercaptoethanol outperformed whole cell yeast and n–3 highly unsaturated fatty acids (HUFA)-enriched yeast as an immune system and growth booster (Kumar, Menanteau-Ledouble, Saleh, and El-Matbouli, 2015).

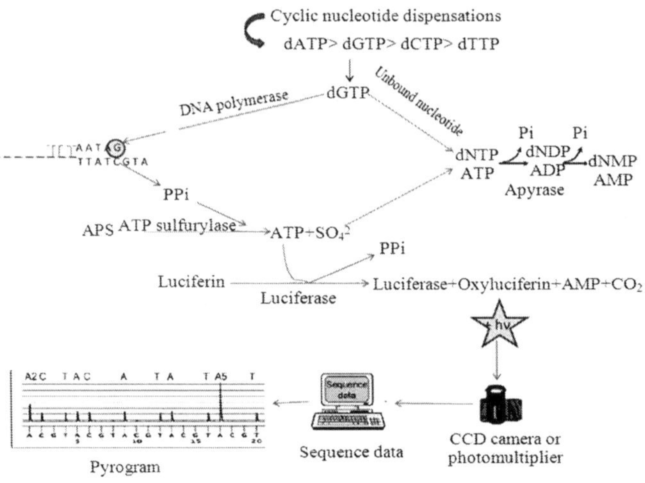

Figure 6.15. Pyrosequencing.

Source: https://www.researchgate.net/profile/Fakruddin-Md/publication/235913524/figure/fig1/AS:341291616489476@1458381670328/Schematic-representation-of-pyrosequencing_W640.jpg.

Aquaculture has grown as an economic activity as a result of global shortages of seafood supplies, primarily in poor nations. Traditional aquaculture has been escalated into reticulated systems with high stocking densities of the farmed species as a result of space and resource restrictions. This creates an artificial environment in aquaculture systems that encourages the growth of pathogenic bacteria and the accumulation of waste metabolites. The uncontrolled release of wasted aquaculture wastes into the environment is also an issue. Disease outbreaks in aquaculture systems, caused by bacterial infections, are a complicated phenomenon linked to stressful environmental conditions such poor water quality, and can result in mass mortality and severe economic loss. The key cause of poor water quality is waste accumulation due to hyper-nutrification, which is caused by high nutrient dietary composition and excessive feeding rates, both of which are typical in intensive aquaculture systems. Fish that accumulate high quantities of nitrogen and phosphorous waste are more susceptible to parasites and diseases, as well as a threat to the environment. Breeders' selection for particular qualities has diminished the vitality of breeding lines in some situations, rendering fish less resilient and more susceptible to disease. The incidence of bacterial disease, which causes damage to fish and frequently results in death, is of special relevance. Gram-negative bacteria like *Aeromonas hydrophila* are among the most common pathogens that cause bacterial disease. Chemicals and antibiotics are commonly used to treat disease, but they affect natural microbial populations, harm the environment, and enhance pathogenic microorganism resistance and virulence.

6.11. DISEASES

Disease is a complex phenomenon that causes the host to suffer some type of measurable damage. Disease outbreaks can either start suddenly and spread quickly, frequently with high mortality rates, and then disappear just as quickly, or they might develop more slowly and with less severity but last longer. Fish disease occurs when the delicate relationship between the hosts, the disease-causing agent, and external factors such as inappropriate environmental changes, inadequate cleanliness, and overcrowding breaks down. A primary invasion by parasites together with stressful environmental factors such as fluctuating temperature and low water quality, is often related with disease outbreaks. Infectious pathogens can cause mass mortality, resulting in severe losses for aquaculture operations. When temperature fluctuations cause immune-modulation, fish disorders such as rotting fins and skin ulcers become more common, resulting in decreased disease resistance and higher mortality. A variety of stress factors in the life cycle of the fish, such as poor water quality, parasite burden, or a natural physiological state are frequently linked to disease outbreaks. Breeders' strict selection for desirable traits has diminished the vigor of breeding lines, making fish less resilient and more prone to disease. Disease is not always caused by the actions of a single bacterial taxon, as disease outbreaks have been linked to representatives of many different bacterial taxa. *Pseudomonas* spp. and *Aer. hydrophila* are two of the most common bacteria that cause fish diseases. Many bacterial pathogens are found in the typical water and/or fish microflora. Not all of these bacteria, however, are primary pathogens; many are opportunistic pathogens that colonize and cause disease in previously damaged hosts. Environmental variables play an important role in disease initiation, which is described as a result of the interplay between the host, environmental stress, and the prevalence of disease-causing chemicals. Some diseases are more common in the spring and are linked to a shift in the environment to warmer temperatures, as well as an increase in the activity of pathogenic bacteria and parasites. Temperature fluctuations produce transitory immunomodulation in fish, which can lead to disease resistance being lowered.

6.12. AMMONIA

Ammonia is a main metabolic waste that is eliminated through bronchial diffusion through the gills. It is also released through sediment mineralization and is created by bacterial ammonification of uneaten food and feces. Through the process of nitrification, ammonia is converted to nitrite and then to nitrate, with ammonia and nitrite being the most hazardous of these metabolites to fish. Nitrite can also be made using the denitrification method. Ammonia concentrations of more than 0.3 mg/l have been found to be toxic to fish, causing hyperplasia of gill tissue (Figure 6.16), gill necrosis, histological signs of kidney and liver damage, and a drop-in growth rate (Naeemi et al., 2013). Epithelial lifting on gill filaments is also caused by high ammonia concentrations, resulting in respiratory impairment and mortality. Because it is a common intermediary in nitrification and denitrification, catabolic ammonification, and nitrate absorption, nitrite is normally present at low quantities in natural systems, unless there is an imbalance. Nitrite can be formed as an intermediary in the conversion of nitrate to nitric oxide, nitrous oxide, and nitrogen gas by denitrification. Nitrite is toxic to fish at concentrations of 0.15 mg/l and higher, as it causes the conversion of hemoglobin to methemoglobin in the blood, resulting in oxygen transport inhibition and mortality from brown blood disease. Dietary phosphorus is an important component of fish feed because it aids in weight gain and feed conversion ratio.

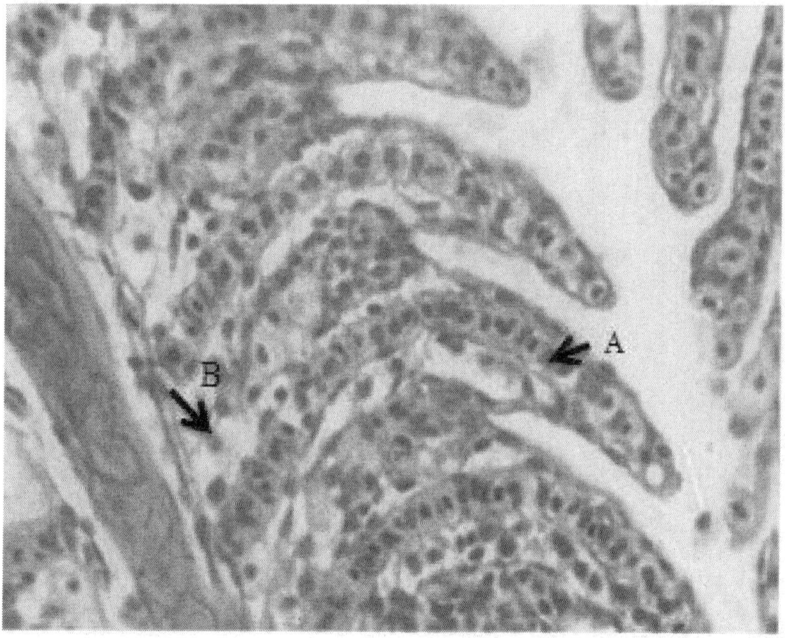

Figure 6.16. Hyperplasia of gill tissue.

Source: https://www.researchgate.net/profile/Milad-Adel/publication/320264717/figure/fig1/AS:546835887595520@1507387243562/Hyperplasia-A-necrosis-and-lifting-of-the-pilar-cell-B-in-gill-tissue-of-rainbow_W640.jpg.

Antimicrobial medicines are widely used to treat disease outbreaks or to prevent disease outbreaks in prophylactic doses. This can result in antibiotic resistance and enhanced pathogenicity, necessitating the use of large doses of existing or new medications to control disease. Antibiotic resistance can endanger human health and cause widespread fish mortality. Compounds employed in aquaculture have also been shown to be hazardous to the fish, with exposure to some chemicals eliciting a stress response and blood biochemical alterations. Higher medication concentrations and a growing variety of chemical residues can have negative consequences for both users and the environment. These compounds also have an adverse effect on aquaculture filtration systems, resulting in a decrease in water quality. Chemicals are often recalcitrant, lasting for days to months and causing changes in naturally occurring bacterial populations. Regulators have recognized the risks posed by chemical use, as evidenced by the ever-growing list of banned substances, resulting in a reduction in aquaculture treatment options. Antibiotic use in animal production has recently been subjected to significantly stricter regulations by governments and organizations. In 1997, the European Union (EU) prohibited the use of avoparcin, and in 1999, virginiamycin, spiramycin, tylosin, and bacitracin were added to the list of prohibited growth promoters in animal feeds (Casewell et al., 2003).

Alternative disease control and water quality enhancement technologies are badly needed due to the challenges in traditional aquaculture practices. Microorganisms have an essential role in aquaculture, particularly in terms of nutrient cycling and animal nutrition, water quality, disease management, and effluent environmental effect. Beneficial microbes can be used to change or regulate the bacterial flora in a water system to improve fish production by lowering pathogen concentrations, improving water quality through waste ion reduction and accelerated mineralization and nitrification, reducing algal growth, and hastening sediment decomposition. These biological agents also offer the benefit of natural integration into existing ecosystems, as well as the possibility of developing multi-effect solutions that are appealing to end consumers. The use of helpful microorganisms rather than chemicals is a more appropriate cure, but its use requires knowledge of the ecological processes that occur in aquaculture systems, disease agents, and the beneficial properties of bacteria to be utilized as biological agents.

Competition with pathogens for adhesion sites, immunological activation, antimicrobial production, competitive exclusion, bioaugmentation, and bioremediation are all mechanisms of probiosis. Although several biological treatments have been created in the previous decade, the approach has largely been empirical, with the exact mechanisms of action rarely revealed, which has hampered technological uptake and application in aquaculture. Competition for attachment sites on gut or other tissue surfaces is one putative strategy for avoiding pathogen colonization. Bacteria must be able to cling to enteric mucus and cell wall surfaces in order to establish themselves in the intestines of fish. Fish infections have been shown to attach to and thrive on or in intestinal or external mucus in *in vitro* conditions. Competition with pathogens for adhesion receptors may be the earliest probiotic impact of a biological agent, because bacterial adhesion is critical during the early stages of pathogenic infection. Immuno-stimulants are chemical compounds that stimulate an animal's immune system and make it more resistant to infection. Fish larvae, shrimps, and other invertebrates have less developed immune systems than their adult counterparts, and their resistance to infection is mostly based on non-specific immune responses. Bacteria may operate as immunostimulants in fish and shrimp, but no solid evidence that they improve the immune response of cultured aquatic species has been found.

Chemical compounds released by microbial populations can have bactericidal or bacteriostatic effects on other microbial populations, causing inter-population relationships to change. The presence of bacteria that produce inhibitory compounds is hypothesized to act as a barrier to opportunistic pathogen proliferation. Antibacterial effects are caused by bacteria producing antibiotics, bacteriocins, siderophores, enzymes, hydrogen peroxide, or changing pH by producing organic acids, ammonia, or diacetyl. Many publications attribute the inhibitory effects shown in *in vitro* antagonism studies to bacteriocins or antibiotics without looking into alternative possibilities. It has been proposed that primary metabolites or merely a reduction in pH can account for growth inhibition in several cases. However, at this point, the link between amensalistic activity and in vivo probiotic activity is fragile at best.

Different microbial populations may coexist in the same ecosystem due to competition for nutrients or available energy, although there have been no extensive investigations on this subject so far. Heterotrophs compete for organic substrates as both carbon and energy sources in aquaculture conditions, which dominate the microbial community. Multiple studies have shown that competitive use of these substrates can reduce target harmful bacteria. In vivo, a bacterial strain selected for its active growth in organic-poor medium was reported to prevent *Vibrio alginolyticus* infection. Because the strain exhibited no in vitro inhibitory impact on the pathogen, it was assumed that it was the result of competitive exclusion. In another case, in vitro antagonism testing revealed no extracellular inhibitory substances, despite the fact that live cells were needed to protect Artemia from pathogenic *V. alginolyticus*. The selected bacteria were thought to protect themselves by contending with the disease for chemicals and available energy.

Iron is required for the growth of almost all microbes. Siderophores are ferric ion-specific chelating agents with a low molecular weight that can dissolve precipitated iron and make it available for microbial growth. The ecological importance of siderophores stems from their ability to scavenge a critical nutrient from the environment while denying competitors access (Ahmed and Holmström, 2014). Many viruses have a high need for iron in low-iron environments, and several investigations have found a link between iron availability and pathogen proliferation. Salmon mortality increased linearly with dietary iron level in a test with pathogenic *V. anguillarum*. Siderophore-producing *P. fluorescens* has been shown to inhibit a variety of Gram-positive and Gram-negative bacteria, especially when iron is scarce. In vitro co-culture studies demonstrated that the filter-sterilized supernatants from iron-limited *P. fluorescens* cultures hindered the development of *V. anguillarum*, but not from iron-deplete cultures. When the culture was treated in vivo with *P. fluorescens*, there was a 46% reduction in mortality owing to *V. anguillarum* infection in rainbow trout juveniles (Verschuere, Rombaut, Sorgeloos, and Verstraete, 2000). It should be highlighted, however, that the body of evidence supporting biological agents' competition for free iron as a method of action is still limited and at best circumstantial.

Farmers can decrease the build-up of dissolved and particulate organic carbon during the culture cycle by maintaining larger numbers of these Gram-positive bacteria in production systems. Nitrite buildup can be produced by nitrate and nitrite reductase activity imbalances, as well as oxygen inhibition of nitrite reductase. Bio-communities, on the other hand, frequently contain bacteria with varying nitrate and nitrite reductase activities, boosting the overall bio-denitrification community's efficiency. Despite the fact that heterotrophic bacteria have lower specific nitrification activity than chemoautotrophs, their total impact on denitrification could be greater due to their higher cell numbers and resistance to process changes. In order to improve water quality in aquaculture, biological agents for nitrification and phosphate bioremediation have value. Many bacterial strains have been found to have an algaecide effect on a variety of micro algal species. Where algal blooms are a concern, generating obstructions in flow systems and variations in oxygen concentration owing to algal cellular respiration, this effect is useful. The combination of pathogen suppression, rapid growth rate, and water quality improvement in bacterial consortia gives a broad spectrum of benefits in a single product.

A mixed Bacillus culture increased the survival and net output of channel catfish in a farm trial, although the route of action was not defined. *Penaeus monodon* larvae fed with Bacillus supplemented Artemia exhibited significantly shorter development times and fewer disease concerns than larvae raised without the Bacillus strain. Based on tests conducted on numerous farms in Indonesia, it was also determined that using Bacillus in penaeid culture ponds increased shrimp productivity by reducing mortality caused by luminous Vibrio spp. Despite the popular perception that autotrophic bacteria dominate the nitrogen removal process, *Bacillus* spp. play a role. *B. subtilis* and *B. cereus*, for example, may grow in aerobic, facultative aerobic, and anaerobic environments, allowing for nitrogen metabolism shifts that facilitate both nitrification and denitrification. *B. subtilis's* nitrite metabolism pattern was shown to be unique among strains that switched between nitrite and nitrate metabolism. The addition of Bacillus species to *C. carpio* culture systems has also been shown to reduce phosphate concentrations. Solubilization improves the bioavailability of bound phosphate, which is expected to promote phosphate removal and minimize the likelihood of algal blooms (Saeid, Prochownik, and Dobrowolska-Iwanek, 2018).

Aquatic animals are high in proteins, trace minerals, and PUFAs, making them nutritionally beneficial for human diet. As the world's population grows, demand for a variety of fish and shellfish products has increased significantly. Because fisheries alone cannot meet consumer demand for aquatic products, aquaculture provides a critical alternative resource. Over the last few decades, aquaculture has grown in importance and intensity, and it is now the fastest growing food production business. Finfish, crustaceans, mollusks, and algal plants are all intensively produced under regulated conditions in modern aquaculture. Aquaculture yields far exceed those of wild fishing, and it provides a reliable source of high-quality seafood and seafood products all year. Aquaculture not only offers local food security, but it also improves people's livelihoods in many underdeveloped coastal areas. Mariculture has grown into a profitable global industry. It has a bright future because of high prices, as well as a worldwide fall in fisheries production due to overfishing and poaching. Abalone is primarily served as a celebratory meal, particularly during weddings and other significant occasions like the Chinese New Year. As demand for live, dried, and canned abalone grows, the already high prices of this marine delicacy continue to rise. The abalone family includes roughly 56 species, all of which are members of the genus Haliotis. Many members of this family have commercial status as fisheries and/or aquaculture species, and are economically significant. Live abalone generates higher income, but transportation and related logistics are difficult to manage. Because of the high demand for this prized seafood, availability is severely limited, resulting in an increase in the number of abalone production facilities around the world.

The Food and Agriculture Organization (FAO) defines probiotics as live microorganisms that give a health benefit to the host when administered in sufficient concentrations. The potential benefits, according to this description, are numerous, and probiotics supplied to shellfish or fish in intensive cultivation could boost their production. Virus and bacterial diseases/infections are well-known to be one of the most pressing issues in aquaculture production today. Probiotics can help with this problem by producing inhibitory compounds like bacteriocins, competing for adhesion sites with opportunistic or pathogenic microorganisms, competing for nutrients with other bacteria, or improving immune status. Several studies have already established the benefits of probiotics in shellfish and fish, although they have mostly focused on their immune response effects. As a result, the potential and intended results of providing probiotics to cultured shellfish or fish will be improved antiviral and antibacterial defenses.

CHAPTER 7

IMPACT OF CLIMATE CHANGE ON FISH FARMING

CONTENTS

7.1. Introduction 152

7.2. Marine Fisheries 152

7.3. Climate Change 155

7.4. Capture Fisheries 158

7.5. Sea Surface Height 159

7.6. Breeding Grounds 162

7.7. Temperature And Salinity 167

7.1. INTRODUCTION

The sun supplies energy and the ocean currents transport the sun's heat throughout the globe. Weather, wind, cyclones, and rain are all fueled by this heat. The sun's irradiance, the earth's albedo, aerosols in the stratosphere, greenhouse gasses in the atmosphere, and ocean currents are the five fundamental climate drivers. Any one of these can cause the others to change, and any alteration results in a change in global temperature. Climate change is caused by temperature changes. Day to night, winter to summer, the temperature changes dramatically at any location on Earth. These variations are all part of local weather and are averaged to produce the local climate, which includes the local average temperature, rainfall, and winds. Obviously, temperature varies with latitude, and the temperature difference between the tropics and the poles is what causes weather patterns. The sun is responsible for the Earth's temperature, which is a manifestation of its heat. When the sun is directly overhead, there is roughly 1.36 KW of solar energy per square meter, which decreases geometrically toward the poles to about 340 w/sq m overall. The tropics absorb a higher amount of the sun's energy than the regions north and south of the tropics, resulting in warm tropical waters and cold Arctic and Southern Oceans.

7.2. MARINE FISHERIES

Marine capture fisheries (Figure 7.1) and aquaculture are vital sources of food and livelihood. Inland aquaculture has grown quickly in recent decades, but marine capture fisheries have plateaued since the early 1990s. In 2013, inland aquaculture produced two-thirds of all cultured food fish, but mariculture or marine aquaculture has expanded as well, from less than 5 million tons per year in 1980 to more than 25 million tons in 2013 (Sorensen and Palomares, 2021). Although there is evidence that climate change has altered fisheries productivity and species distribution at regional and local scales, there is no clear evidence that climate change has affected total global production to date. Given their importance as a key component of worldwide human food consumption, fisheries, along with agriculture and other food sources, must be included when assessing future food security. When it comes to detecting climate effects, attributing changes to specific elements, and predicting the repercussions of future change, however, marine capture fisheries differ significantly from agriculture. In contrast to agriculture, marine fish are taken from natural ecosystems with no control over inputs like fertilizers, irrigation, predators, pests, or diseases.

Figure 7.1. Marine capture fisheries.

Source: https://www.financialexpress.com/wp-content/uploads/2018/02/Fishermen-R-L-1-620x400.jpg.

After four decades of rapid growth, global capture fish production remained steady in the 1990s at roughly 80 million tons per year, a level that has remained consistent into the 21st century. The growth in worldwide production is due to the exploitation of previously unexploited areas and stocks, as well as higher yields from previously exploited stocks. Many exploited stocks got overfished, resulting in lower yields and significant consequences for world output. Overfishing has resulted in the closure of some major fisheries and a reduction in productivity in others, but recent stricter management measures in many ocean areas have allowed major stocks to rebuild and fisheries to resume. The relative significance of these competing forces to worldwide capture production trends are still debated. Global fisheries production is unlikely to increase unless overfished stocks are allowed to recover and all stocks are fished at levels that maximize sustainable yields. There is hopeful evidence of small progress in global capture fisheries management, and a little faster pace of fisheries reform might bring the status near to ideal by 2050 (Garcia and Grainger, 2005). The areas with the highest capture output include the Northwest, Western Central and Southeast Pacific, and the Northeast Atlantic. China is by far the greatest fish-capture-producing country, accounting for more than half of the global catch, along with Peru, Indonesia, the United States, the Russian Federation, Japan, and India. Fish consumption per capita has increased three-fold since 1950, but it varies by region, with the highest consumers approaching or exceeding 100 kg per person per year, while the lowest consumers in Africa barely reach 1 kg per person per year.

Millions of people rely on capture fisheries and aquaculture for a living, with an estimated 120 million people employed full or part time in fisheries and over 19 million in inland and marine aquaculture combined. Habitat degradation, overfishing, eutrophication, pollution, and species introductions have all had an impact on fisheries, sometimes to the point of extinction. These anthropogenic threats to future production and resilience are well understood, and management measures to fight them are in place or being created in many regions. Fish resources in Europe's rivers, estuaries, and coastal areas began to decrease centuries ago, long before motorized fishing methods were created, due to increased demand from a growing human population. To supply European markets, distant water fisheries were extending to Iceland and America by the 16th century. Demand for fish, combined with short-term social and economic pressure, leads to overfishing, which is still the main hindrance to successful management. Since 1971, the seas have held more than 90% of the energy created by global climate change. As a result, the surface ocean has risen 0.11°C each decade over the same time span. The surface of the ocean is expected to warm the most in the future in tropical and Northern Hemisphere subtropical regions, as well as at greater depth in the Southern Ocean (Levitus et al., 2009). Ocean acidification is one of the more visible impacts of rising CO_2 emissions in the atmosphere. Over a quarter of anthropogenic CO_2 emissions have dissolved and entered the ocean, altering seawater chemistry. The pH of the ocean surface has dropped by 0.1 units since the industrial era, which, while minute in comparison to natural pH variations in coastal areas, is likely to have an impact on ocean ecology. Depending on greenhouse gas emissions, projections through the end of the 21st century imply significant acidification and a pH fall of 0.06 to 0.32 units.

Warming, acidification, and deoxygenation produced by warming of the ocean surface will affect organisms and ecosystems. Warmer water carries less oxygen in solution, and, more critically, surface warming inhibits vertical mixing and therefore O_2 transfer to the ocean's deeper layers. Under the worst emission scenarios considered by the IPCC, open ocean primary output might drop by up to 9% globally by 2100. This is due to predicted increases in high-latitude production and decreases in temperate and tropical regions. It's impossible to say how changes in the global ocean's physicochemical features, as well as biological production, would disturb ocean ecosystems. Warming will trigger species-specific responses, such as increased metabolic rates and increased heat tolerance, which will have varying effects on abundance, distribution, and community structure. Many species of invertebrates and fish, as well as marine mammals, have already changed their distribution and seasonal patterns as a result of warming.

7.3. CLIMATE CHANGE

Climate-related parameters, especially temperature, have a direct impact on biochemical reaction rates, which govern cellular activities and metabolic function like respiration and absorption. Individual and population fitness are altered as a result of these changes in fish growth, reproduction, and behavior in the wild and in aquaculture. Although we are primarily interested in the effects of climate-related changes on large scales, knowing common molecular and cellular pathways is critical to our capacity to model these scales. Ecology and physiology are linked scientific fields for understanding climate impacts. Ecology deals with the dynamics from individuals up through populations to communities and ecosystems, while physiology deals with the dynamics from people down to organs, cells, and molecular processes. Fish production is dependent on primary production, which is influenced by direct temperature impacts on phytoplankton production and remineralization, as well as indirect temperature effects due to changes in nutrient supply and light environment due to vertical mixing variations. The net impacts of increasing temperatures on phytoplankton and macroalgae are unknown, adding to the uncertainty in primary output forecasts. By statistical comparison of observed regional and temporal patterns, the relationship between climate parameters and the distribution and rates of increase of different species of fish may be explored and predicted empirically. It can also be modeled, and a combination of statistical and mechanistic models is likely to make the best use of data and produce the most accurate estimates.

By 2050, the highly resolved shelf model predicts a 24% increase in potential fisheries production in West Africa, but a bioclimate envelope model combined with the first published GCM (Figure 7.2) estimates of global primary production predicted a 26% decline in potential fisheries production for the same period and area under the same climate scenario

(Barange et al., 2014). Although the former estimate is more reliable because the model on which it is based includes important coastal and shelf processes, the disparity between these projections highlights the difficulty of estimating ecosystem production in regions where local conditions may predominate over global patterns. Neither of the projections provides a complete and comprehensive description of the underlying assumptions and uncertainties, which can be used to assess their reliability; this is a typical defect in complicated, multistage models, as is the experts' overconfidence in their results. Climate change is expected to cause changes on a lower scale than those that have occurred naturally in the past in most locations due to climate variability. Climate change will have an impact on food system sustainability, future catch composition, and the costs of producing fish for food, but it will not be the only the most important factor in achieving the required levels of fish production to feed a growing, wealthier global population with a higher fish protein intake. Better fisheries management is predicted to result in sustained and improved fisheries production in most regions, overcoming potential climate change negative effects at least until 2050 (Free et al., 2020).

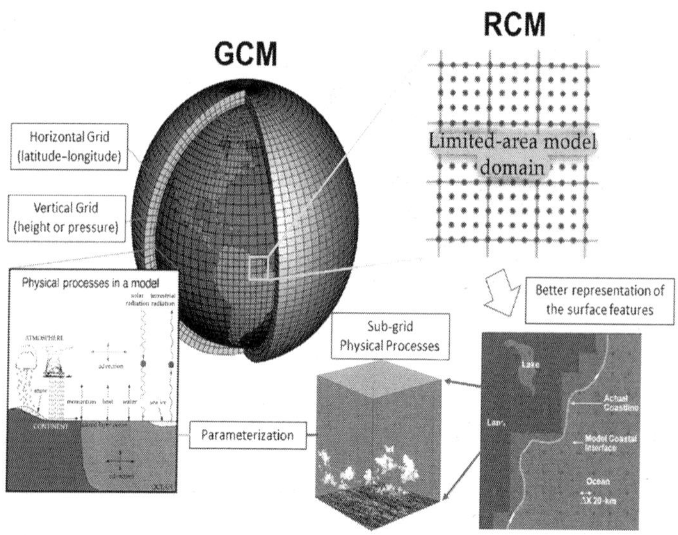

Figure 7.2. GCM.

Source: https://www.researchgate.net/profile/Marta-Llopart/publication/326711713/figure/fig1/AS:718029935632385@1548203085584/Illustration-of-a-GCM-RCM-physical-processes-in-a-model-and-refinement-of-coastline_W640.jpg.

The Baltic Sea serves as an example of the errors that emerge when GCMs are linked to regional climatic and biological models, but it also demonstrates the utility of scenario-based risk assessment for fisheries strategic management. Regionally, processes that are poorly represented in GCMs may be dominating. With salinities varying from totally oceanic to freshwater, the Baltic Sea is the world's largest confined sea. Salinity has a higher impact on the biological response of fish in the Baltic Sea than temperature, and whether or not cod, the most important species, will be able to breed relies heavily on whether and how much salinity drops. However, depending on which global model is chosen to impose the regional scenario, the expected changes in salinity vary significantly. The pattern and strength of wind forcing, as well as the volume of precipitation, are crucial for Baltic Sea salinity, but they are not accurately reflected in GCMs. Despite the fact that the likelihood of a salinity decline cannot be quantified using current climate models, the consequences of a range of salinity changes can be projected using a stochastic food web model that shows how future salinity would alter the risk that fish stocks biomass would drop below acceptable levels at given fishing levels.

Ocean warming, sea level rise, ocean acidification, weather pattern changes, and extreme weather events are the key direct and indirect drivers of climate change impacts on agriculture. Temperature influences growth rates, disease susceptibility, spawning timing, and mortality at different stages of the life cycle, as well as having economic implications due to direct impacts on the culture process. Submergence, coastal flooding, coastal erosion, and saltwater intrusion will become more common in coastal systems and low-lying locations. Bivalves and some crustaceans may be affected by predicted ocean acidification in terms of growth, reproduction, and shell quality. Bivalve larvae have already died in large numbers as a result of acidity. Bivalve production accounts for a large component of global mariculture, and coastal water acidification could have a big impact on the industry, particularly in Asia. Warming could benefit some farmed species by enhancing larvae settlement, attachment of young bivalves to the substrate, and growth rates, as well as extending their latitudinal range. Acidification's effects on marine finfish are unknown. Acidification affects fish embryos and larvae more than juveniles and adults, and there may be sublethal effects such as slowed growth rates.

7.4. CAPTURE FISHERIES

Capture fisheries and aquaculture have varied operations, and management, and are normally viewed as separate sectors. However, there are connections and overlap between the two sectors that may need to be considered when dealing with climate change implications. Culture-based fisheries, for example, that involve stock improvement of naturally occurring species, include both sectors and are becoming increasingly important. When done correctly, stock improvement can be a low-cost and ecologically friendly way to increase food production. Aquaculture is sometimes marketed as a means of mitigating the negative effects of climate change on fish catches. Aquaculture's global expansion is already improving food supply and helping to offset decreases in wild fisheries productivity due to overfishing or future climate consequences. Aquaculture, on the other hand, can have a negative impact on fisheries, as seen by the vast-scale conversion of mangroves into fish and shrimp ponds in Southeast Asia, which has caused coastal erosion and put large areas at risk of floods. As sea levels rise owing to climate change, the problem is only going to become worse. Aquaculture expansion must be handled properly to ensure that new development is climate change resistant and does not have detrimental consequences for marine ecosystems and the resources they support.

Although both are meant to assist us in making future decisions, and both employ a combination of mechanistic and statistical modeling, projections of future world fish output differ from the type of weather forecast. A weather forecast informs you what to anticipate tomorrow, typically in images, however a forecast of future fish output is generally for a longer period of time and includes a variety of factors. It is quite straightforward to assess the accuracy of a weather forecast, but with a fisheries projection, you must examine what occurred to all of the factors, which may be difficult or impossible. The accuracy of a projection is determined by the data and model utilized, as well as the uncertainties. The IPCC, for example, predicted changes in species richness and fishery catch potential at high and mid-latitudes by 2050, but the variables included ocean acidification, oxygen, and trophic structures, all of which are likely to be affected by climate change. More variables are being included in the models, allowing their contribution to the projections and the estimated uncertainty to be taken into account. The fisheries estimate also starts with the assumption that a specific climatic scenario would occur, which adds to the uncertainty. Wherever possible, predictions for each scenario should be made, demonstrating the range of possible future outcomes in the projections. Because we cannot do trials that control for the many elements that may have an effect like food availability, predators, pests, pH, dissolved O_2, fisheries estimates are inherently considerably more unpredictable than agricultural projections. Unlike monocultures of a few plant crops that produce much of the terrestrial food supply, marine systems are relatively diversified, and we catch fish from numerous trophic levels along the food chain. Because regional and local projections of fisheries production are closer to the scales at which fishing enterprises and regulatory systems operate, there is a lot that can be done to make them more reliable, including improved regional climate projections, inclusion of coastal and shelf sea processes, coupled models of ocean dynamics, biogeochemistry, lower trophic levels, and fish populations reconstruction and assimilation.

Because of the biological, technological, social, and economic variability of fisheries, no uniform recipes for fisheries adaptation exist, and each case must be assessed separately. Nonetheless, interventions that decrease the effects of all known existing pressures on marine ecosystems typically improve adaptation to future stresses caused by climate change. Because their demographic and taxonomic diversity is lost, disturbed, and stressed ecosystems are less resilient to the effects of climate change than healthy ecosystems. Terrestrial agriculture and aquaculture can adjust to climate change by planting new crops, changing planting and harvesting seasons, changing irrigation, and so on, but such adaptability is unlikely for capturing fisheries that gather a variety of species from natural systems. Changes in fishing methods, fisheries infrastructure, and markets are the main ways we may adjust catch fisheries to climate change, although there may also be some natural acclimatization and adaptability within fish populations. Fish acclimation to conditions encountered during their lifetime and different proportions of polymorphic forms within a population are feasible, but evolutionary processes move at a slower pace than climatic change. Research showed a variety of feasible activities and supporting policies for fisheries and aquaculture adaptation to climate change in the tropical Pacific.

7.5. SEA SURFACE HEIGHT

Sea surface height (Figure 7.3) or SSH can be affected by ocean warming, with increased surface heights coming from water expansion as it heats. Satellite altimetry observations around New Zealand reveal that SSH grew from the 1990s to 1999, then marginally decreased. Since the late 1800s, coastal tide gages have been used to record the sea level in New Zealand.

Long-term measurements from New Zealand's four main Standard Ports, as well as six additional sites, show a 1.7 mm/y increase in sea level relative to the adjacent land mass from 1900 to 2008, which is consistent with global sea-level trends (Church and White, 2011).

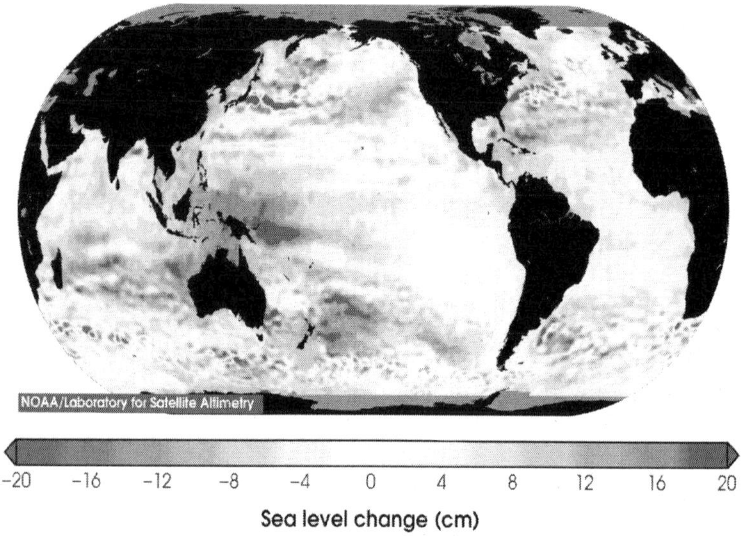

Figure 7.3. Sea surface height.

Source: https://d32ogoqmya1dw8.cloudfront.net/images/getsi/teaching_materials/climate_change/climate_chane_unit_2_image.png_1536.png.

Further work to understand this New Zealand tidal gage data is ongoing, as variations in sea level from coastal tide gages are affected by local tectonic motion. Precipitation, air and water temperature, and surface winds are expected to be affected by changes in atmospheric and oceanographic conditions, as well as ocean salinity and stratification. Changes in ocean stratification and currents affect concentrations of important nutrients like nitrate, phosphate, and silicate in surface waters, which can affect the type of phytoplankton present, their seasonal cycles, and overall rates of primary productivity, which can affect food webs and fisheries. Satellite measurements of ocean color have been used to investigate spatial patterns in the concentration of chlorophyll a or Chl-a around New Zealand as a representation for primary production by phytoplankton. The Subtropical Front east and west of New Zealand has higher chl-a concentrations, while northern Subtropical waters and southern Subantarctic waters have lower values. Measurements of ocean color from two recent satellite sensors like

SeaWiFS and MODISAqua were used to construct an 18+ year timeseries of chl-a in New Zealand's oceans texplore temporal variations in ocean productivity. Monthly chl-a anomalies were typically small at the scale of the New Zealand EEZ (Blondeau-Patissier et al., 2014).

Water temperature, coastal wave climate, sea level, river flow, acidification, coastal primary productivity, and coastal connectivity as influenced by water structure and oceanographic currents are all likely to change in New Zealand coastal regions as a result of climate variability and change. Information on species tolerances to environmental variability is scarce in New Zealand. It's also difficult to predict how changes to individual species will affect communities. Some estuaries, wetlands, and seagrass beds in New Zealand may be particularly vulnerable to climate change and these often serve as important nursery habitats for commercially and recreationally important fishes and invertebrates. Mangroves have been expanding in some places of New Zealand, despite recent losses of seagrass and salt marsh habitat due to reclamation and draining of coastal land. For example, mangrove coverage in New Zealand's Hauraki Gulf region has been reported to be rising at a rate of about 4% per year, owing to increased coastal sedimentation but also possibly due to the influence of warmer seas.

Rock intertidal platforms and shallow subtidal reefs are two other key coastal habitats in New Zealand. The effect of extreme wave events on seaweeds, which provide food resources, nutrient recycling, and physical habitats for fish and invertebrates in New Zealand rocky reef ecosystems, makes these coastal habitats vulnerable to changes in wave conditions associated with climate change. Because of the strong relationship between some harvested species and macroalgae, improved information on coastal changes is critical, and research is undertaken in New Zealand on topics including macroalgae distribution shifts. In New Zealand, there is also continuing study on the effects of environmental change on fish, with a particular focus on the sublethal effects of ocean acidification and the potential ramifications for New Zealand fish and fisheries. Urchins, lobsters, bivalves, gastropods, and coralline algae may be particularly vulnerable to acidification in New Zealand's coastal areas. For example, under higher dissolved CO_2, there is clear evidence of malformation of sea urchin larvae in tropical to Antarctic species, including those from New Zealand; this may result in smaller larvae and a longer planktonic phase, reducing the chances of survival to the adult stage. Kina or *Evechinus chloroticus*, an endemic New Zealand Sea urchin that is commercially profitable and considered a delicacy by many, is a source of concern in New Zealand.

Climate change may also have an impact on paua or Haliotis iris, an endemic New Zealand abalone with a commercial fishery worth NZ $50 million and a huge recreational fishery (McShane, Mercer, and Naylor, 1994). Adult paua appear to be more resistant to changes in temperature and acidification than juvenile paua. Experiments on the effects of acidification on juvenile paua in New Zealand waters revealed that while survival was unaffected, growth was greatly reduced, and shell breakdown was visible. Flat oysters' development and shell surfaces showed similar impacts. Paua fishermen have reported more storms, isolated brown algal damage, increased sedimentation, and the effects of acidity on paua recruitment. Many important fisheries species have planktonic life phases in the ocean. Physical oceanography that includes water temperature, salinity, distribution of water current, etc., associated with climate change has a significant impact on plankton ecology. Fish and shellfish larvae are crucial natural seeds for stock improvement and aquaculture, and knowledge of the ecology of early life stages of aquaculture-relevant species is essential for the development of aquaculture systems. During their early stages, certain important species for Japanese fisheries such as yellowtail, tuna, and eel live in and around the Kuroshio warm current. As a result, physical oceanographic effects, and other aspects may be critical for their survival. Their survival has an impact on stock size and seed catch for aquaculture, so climate change is a concern not only for natural fisheries, but also for aquaculture success.

7.6. BREEDING GROUNDS

The breeding grounds of five-ray yellowtail (Figure 7.4) or *Seriola quinqueradiata* (Figure 7.5), bluefin tuna or *Thunnus orientalis*, and Japanese eel or *Anguilla japonica* occur in the warm current. They are extremely valuable species in Japanese aquaculture. Aquaculture seed has been harvested from young fish collected in natural habitats. The distribution and survival of these species' early life stages have an impact on aquaculture's success. Global warming may result in increased current speed of the Kuroshio and Tsushima warm currents, as well as higher water temperatures. However, no accurate findings exist that provide predictions about the ecology of these species' larval stages as a result of warming circumstances. Because of a scarcity of food or their own physiological state, the high temperatures may not be ideal. Both the Kuroshio and

Tsushima warm currents are anticipated to accelerate up as a result of global warming. Faster current speeds may help the larvae get from the upper to the lower stream faster. Because the spawning grounds of yellowtail and bluefin tuna are not far from the Japanese archipelago, larval fish may pass through Japan's food-rich coastal area before growing large enough to survive in waters of Japan. The breeding grounds of the Japanese eel, on the other hand, are distant from Japan, in the middle of the North Pacific around the Mariana Trench, and many larval eels remain along the eastern Chinese coast, downstream of the Kuroshio warm current off Japan. If larval eels can successfully live under global warming conditions, they may be more likely to remain around the Japanese coast rather than the Chinese coast, off the upper stream of the Kuroshio Current.

Figure 7.4. Five-ray yellowtail.

Source: https://upload.wikimedia.org/wikipedia/commons/thumb/f/fe/Yellowtail_snapper_seen_from_Underwater_Tunnel_Atlantis.jpg/800px-Yellowtail_snapper_seen_from_Underwater_Tunnel_Atlantis.jpg.

Figure 7.5. Seriola quinqueradiata.

Source: https://upload.wikimedia.org/wikipedia/commons/a/a9/Seriola_quinqueradiata_%28200810%29.jpg.

Spiny lobsters (Figure 7.6) spawn in the coastal area of southern Japan, where the Kuroshio warm current meets the northern boundary. Spiny lobster larvae develop in the summer on rocky coastal areas and reach the Kuroshio warm current, which transports them eastward. Then, by vertical migration behavior, some of the successful larvae can shift to Kuroshio's outer counter current and migrate southward. After about a year of growth, the post larvae re-join the Kuroshio warm current and transfer to the inner counter current, eventually reaching the southern Japanese coast and settling on the rocky seabed. Japanese spiny lobster requires this cyclical movement to replenish their stock. The distribution and current speed of the Kuroshio warm current, as well as the timing and depth of their vertical migration, are all factors in the spiny lobster's survival rate. The eastern Australian region spans 3,860 kilometers from north to south on the eastern seaboard, with approximately 1,900 kilometers in the Gulf of Carpentaria in the north and another 4,800 kilometers in the south, which includes Tasmania and Victoria (Colberg, McInnes, O'Grady, and Hoeke, n.d.). A large amount of Australia's fisheries resources is supported by this huge coastline. The climate zones in the area range from warm tropical in the north to cool temperate in the south. In recent years, eastern Australia has seen some of the world's most severe environmental changes connected to global warming, particularly in the south-east region. The south-east maritime seas of Australia have warmed at about four times the global average rate. Longer and greater incursions of the warm, nutrient-poor East Australian Current (EAC) into waters off south-eastern Tasmania now occur, bringing both heat and larvae and adults of northern species southward to Tasmania. Compared to the sub-Antarctic water masses that dominated the oceanographic signature on Tasmania's east coast until recently, the EAC is nutrient sparse. As a result, marine systems in southeast Australia are experiencing significant temperature rises as well as nutrient reductions, resulting in a decrease in primary productivity. Significant coral bleaching events on the Great Barrier Reef due to extremely warm sea surface temperature (SST) connected to severe El Nino cycles have also occurred in recent years. The most recent bleaching incident, which occurred in the summer of 2016, was the most severe yet.

Figure 7.6. Spiny lobster.

Source: https://upload.wikimedia.org/wikipedia/commons/thumb/3/3e/California_spiny_lobster.JPG/800px-California_spiny_lobster.JPG.

When attempting to predict how these resources will adapt to future climate change, it is critical to understand the observed local climate and ecosystem responses to climatic variability. Summer monsoon SSTs in north-eastern Australia range from 25°–30.5°C, with 30°–32°C in the Gulf of Carpentaria, whereas winter SSTs in these regions range from 20°–26°C and approximately 26°C. The yearly range of SST in the northern tropics is about 4°C and in the southern tropics is about 6°C, but localized fluctuations can be much greater (Khan et al., 2021). Tolerances to SST are anticipated to be a major determinant of future population distributions, but they also influence spawning and early growth time. Australia has a significant degree of interannual and decadal rainfall variability, making understanding the importance of rainfall extremes and changes in averages difficult to detect. Rainfall occurs on just 30% of days on the east coast in the summer and only 14% of days in the winter. The Gulf of Carpentaria is located in the wet-dry tropics and receives a lot of rain. Weipa, for example, receives 1,768 mm of rain annually, the most of which falls between November and April. Northern Australia's highly seasonal and variable rainfall regime also results in very fluctuating river flows. Between 17° and 23°S, the majority of total river flow into north-eastern coastal waters occurs, with the highest annual flow in March. Australian river flows are among the most variable in the world due to high rainfall variability. River flow has been shown to be a significant determinant of early survival, growth, and capture rates for a number of important nearshore fisheries species in northern Australia.

Barramundi is a big teleost found throughout the Indo-Pacific region's tropical and subtropical waters, including northern Australia. They can be found in a variety of environments, including coastal foreshores, estuaries, tidal creeks, swamps, flood plains, coastal lagoons, and upstream rivers where the sea is accessible. They are an iconic and popular target species of commercial, recreational, and indigenous fisheries in northern Australia, having significant economic value for commercial fishing and tourism in particular. Barramundi have a complicated and spatially varied life history, such as migration from freshwater to saltwater to spawn. Female barramundi is thought to live in the lower regions of estuaries and along the shoreline. Male and female barramundi are often separated by their life cycle, with smaller and younger male fish dwelling in the upper estuary or freshwater parts of the river. Tropical rock lobster fishing is only done in

NE Queensland and the Torres Strait, and the ornate rock lobster is the most common species caught. Commercial collection methods include hookah or free diving and hand collection, nooses (snares), or hand spears. Divers operate nearly exclusively during daylight hours around coral reefs at depths of up to 20 meters. Despite the poor output from the fishery, the product has a high value and is especially vital economically for the Torres Strait's traditional owners. Although there is a tiny domestic market for the product, the majority is sold to mainland China via Hong Kong as frozen tails or live tails. The distribution of *P. ornatus* phyllosomas and pueruli in relation to ocean currents in north-eastern Australia supports the hypothesis that phyllosomas are transported from the Gulf of Papua breeding grounds into the Coral Sea Gyre and then onto the Queensland coast, Torres Strait, and Southeastern Papua New Guinea by the Hiri boundary current.

7.7. TEMPERATURE AND SALINITY

Temperature and salinity have a considerable impact on juvenile *P. ornatus* development and survival, with maximum growth in 25°–31°C water and lowest survival but fastest growth at 35 ppt (Vidya and Joseph, 2012). Furthermore, professional divers in Torres Strait observed that *P. ornatus* went deeper into cooler water during recent hot years, making them less accessible for collection by the regularly employed free diving approach. The location and intensity of recruitment of *P. ornatus* throughout the fisheries grounds are highly influenced by oceanic currents in the north-western Coral Sea. Warming waters and probable changes in Coral Sea current patterns are anticipated to have the largest impact on *P. ornatus*. Although recent studies of the sensitivity of tropical rock lobster to climate change found that they were moderately vulnerable overall, they were rated one of the highest priority species for further research and possible adaptation measures due to their local fishing importance. Their apparent high susceptibility to rises in SST is particularly concerning. If the number of hot years continues to rise, *P. ornatus* catchability may become more variable, resulting in economic ramifications and less confidence for fishermen. They may possibly begin to expand their range southward. Increased SST is also expected to alter early development, with faster growth and increased larval supply, but it may potentially reduce juvenile survival, with unknown population repercussions. Climate change's impact on ocean circulation in the western Coral Sea is one of the critical factors that could have an impact on *P. ornatus* populations. Given the importance of this current in larval movement and the length of oceanic larval development, recruitment to Australian waters could be severely hampered. If the following effects materialize, early warning signs could include increasing yearly recruitment variability.

Spanish mackerel (Figure 7.7), also known as narrow barred Spanish mackerel, is a fast developing and highly mobile epipelagic fish found in tropical and subtropical waters across northern Australia, mainly along reefs. They can grow to be quite huge, with record specimens being over 2 meters in length and weighing over 100 kg (Chen, 2017). Spanish mackerel fishing is done commercially and recreationally in northern Australia, from the east coast to the west coast, utilizing line fishing trolling tactics with lures or baits. Separately managed fisheries in the Gulf of Carpentaria, the Torres Strait, and the east coast of Australia are based on a metapopulation stock structure. The product is mostly distributed to domestic markets for the fish and chip industry. Fish are caught throughout the year in the Gulf of Carpentaria, although the Torres Strait fishery is seasonal between August and March, focusing on fish near Bramble Cay in the north-western Torres Strait. Traditional people of the Torres Strait value the fishery both economically and culturally. Fish are mostly available throughout the summer/autumn months during a southerly large-scale migration assumed to be linked to seasonal warmer currents flowing southwards. The fishery in Queensland is open all year, although the majority of the annual catch has traditionally come from a recognized spawning zone near Townsville between September and December. Although temperature has been proposed to greatly influence spawning seasonality and annual migrations, the susceptibility of Spanish mackerel to changes in the environment is poorly understood. The increasing strength of the EAC may have influenced the yearly southerly migrations of Spanish mackerel into NSW, although due to low commercial catches and no recreational catch data available, this is not apparent from a superficial study of NSW catch data. With expected future ocean warming and sustained EAC strengthening, Spanish mackerel could become a significantly more important species in NSW, with the prospect of increased southerly migratory and/or increased presence due to range shifting.

Figure 7.7. Spanish mackerel.

Source: https://www.seapedia.net/wp-content/uploads/2021/12/16063130e904b622.png.

Changes in air circulation are linked to recent cold season dryness over southern Australia and evidence of enhanced rainfall over the Southern Ocean. While natural variability is likely to have a part, a number of studies imply that ozone depletion and global warming are influencing circulation and pressure variations, with the southwest bearing the brunt of the effects. Oceans are the world's primary climate-change buffer, absorbing up to 80% of the heat and 50% of the carbon released into the atmosphere ("The Ocean and Climate Change," 2017). Temperature, environmental fluxes, ocean pH, sea level, and wind regimes all have an impact on fish productivity and distribution, with consequences for the communities and fisheries that rely on them. South east Australia's pace of ocean warming is 2.3° higher than the global norm, with nearshore seas rising at more than three times the global average. Increased salt, sea levels, and currents in the southeast are associated with lower pH and rainfall, as well as more frequent extreme events. The EAC, a warm, poleward-flowing seasonal current, has a significant impact on the eastern and south east Australian coastlines. While future changes are difficult to anticipate, climate change has already had an impact on these natural connections. As the climate continues to warm, these tendencies are likely to persist, if not intensify, over the next 50 years. Increased acidity is also expected to diminish the availability of crustose coralline substrates for larval settling and early development. All of this would have a significant influence on recruitment and production. Over the last two decades, alterations in nearshore community structure are considered to have been initiated by the southward and westward strengthening of the generally oligotrophic EAC into inshore waters in eastern Victoria and Tasmania. This is due to the spread of species like the Black Sea Urchin or *Centrostephanus rodgersii* from New South Wales to Tasmania and Victoria, as well as a large decrease in Giant Kelp or *Macrocystis pyrifera* biomass. The extension of the urchin range has resulted in localized abalone population loss and a diminution in abalone habitat. It's difficult to say how climate change may affect Australian abalone stocks. However, climate change is anticipated to affect these abalone stocks, as well as the fisheries that rely on them. Summer SST predictions were very varied across spatial areas and unrelated to historic catch.

Snapper is a huge, long-lived finfish with a broad Australian distribution that extends southwards from Western Australia's mid coast, the southern continental coastline and north coast of Tasmania, and the east coast up to north Queensland. Snapper can be found in a variety of coastal settings, including bays, inlets, gulfs, and open marine seas to a depth of at least 200 meters, across their range. Snapper is exposed to a variety of environmental situations throughout their life cycle. Three separate snapper stocks can be found in Australia's south-east. The Western Victorian stock is thought to expand westward from Wilsons Promontory into South Australian waters around the Murray River's mouth. The South Australian stock, which comprises the populations of Spencer Gulf and Gulf St. Vincent, extends westward from the Murray mouth into Western Australian seas. During the reproductive season, spawning aggregations of highly fertile adult snapper gather, producing huge numbers of small, spherical, buoyant eggs. After 1–2 days of growth, the fertilized eggs hatch, releasing little, underdeveloped larvae measuring 1.5–2.5 mm in length. Because the eggs and larvae are pelagic and free-floating members of the planktonic community, they are prone to passive current transfer. Before settling into nursery regions in bays, gulfs, and estuaries that support either bare, flat muddy habitat or seagrass meadows, the larvae take around 20–30 days to reach settlement competency.

The available data in the Southeast United States Atlantic show that bottom water temperatures have stayed steady and surface water temperatures may have fallen marginally during the previous several decades. Detecting changes can be difficult since data must be collected over wide areas and over lengthy time scales in order to discern sometimes tiny long-term changes amidst substantial natural variability. This is especially true when it comes to detecting climate change in more remote habitats, such as the pelagic fisheries of Hawaii in the North Pacific. Furthermore, individual climate-related changes do not occur in isolation. There are multiple physical, chemical, and biological interactions to consider, and pressures on the system originate from a range of sources, including climate fluctuation and change, as well as rising direct human impacts due to population expansion. With over 160 million people living in coastal communities and fisheries and aquaculture contributing significantly to the nation's social and economic well-being, good stewardship of our living marine resources requires a strong scientific understanding of how the ecosystem is changing, the mechanisms of change, and the ability to anticipate and account for future change (Cohen et al., 2019). Fishing is the most serious threat to the ocean ecology, whereas vessel traffic, mining, and forestry are considered minor threats. Climate change will likely add to the pressures, which will include rising ocean temperatures, rising sea level, changes in circulation patterns, rising CO_2 concentrations, and lowering pH. These physical and chemical changes to the environment may have negative consequences for fisheries, affecting the ocean's economic and cultural benefits, especially for rural areas.

In comparison to other Gulf of Alaska fisheries on non-calcifying animals like finfish, those mediated by ocean acidification are projected to have the most direct impact on shellfish fisheries. The Tanner and Red King crabs have negative reactions to high CO_2 and low pH water. Although short-term effects of ocean acidification were masked by variability in juvenile recruitment, expected yields were estimated to fall over the long term when projections were done with scenarios that included predicted future ocean acidification levels. Assessing risk as a mix of hazard or ocean acidification levels, vulnerability or sensitivity and adaptive capacity, and exposure or economic and nutritional is required to expand the potential economic effect analysis of ocean acidification to human communities. Based on these three factors, a risk assessment of Alaskan communities found that those who rely on ocean acidification-susceptible fisheries species for nutrition and economics, as well as those who live in areas where chemical conditions are expected to change quickly, are among those most at risk of negative consequences. Ocean acidification-mediated climate change effects are most likely to have an indirect impact on Gulf of Alaska groundfish fisheries through food web interactions. Under elevated CO_2 laboratory conditions, there is limited evidence of direct impacts. More study has been done on the consequences of large-scale climate change on groundfish, indicating the far-reaching implications of climate change on marine communities in the Gulf of Alaska.

The effects of ocean acidification on Pacific salmon in the Gulf of Alaska will most likely be mediated through their food source, similar to groundfish. Because pteropods are a key food source for young pink salmon, if pteropod abundances diminish owing to calcification consequences, salmon growth rates are expected to decline. Furthermore, large-scale patterns in salmon productivity have been connected to climate patterns such as the Pacific Decadal Oscillation (PDO), a climate index based on North Pacific SST fluctuation. Diet is one possibility, though the actual mechanisms relating climatic changes to salmon production are generally unknown. During major El Nino and La Nina occurrences, for example, salmon diets in the central Gulf of Alaska, which are generally dominated by squid, had far fewer squid. Climate-induced range shifts may cause changes in the food web, introducing new predator–prey pairings. The world's marine tropical coastal region has a diverse range of environments, human influences, and fisheries resources. Many tropical countries rely on tropical estuaries, coasts, and coral reefs for subsistence fishing. Coastal and estuarine tropical settings support a diverse range of fish species, including recreational and commercially important species. With a high degree of urbanization and agricultural growth, these habitats are frequently under direct human pressure internationally. While some fish species live in estuaries all year, many others, particularly coastal, coral reef, and diadromous species, use them for breeding and nursery grounds.

Reduced freshwater flow and sea level rise, for example, will be particularly critical in low-lying coastal wetlands and mangroves, and may cut off larval and adult dispersal paths. Reduced runoff may cause the salinity of nearshore waterways to rise. Rising sea levels will also affect fish communication between nearshore habitats. At lower latitudes, mangrove systems are more diversified and provide complex habitat for all phases of the fish life cycle. Mangrove environments, for example, provide nurseries for recreational and commercially important species like barramundi, *Lates calcarifer*, and mangrove jack, *Lutjanus argentimaculatus*. Climate change puts tropical seagrass systems at risk, which is compounded by their closeness to farmland and urban areas, as well as risks such as excessive turbidity, dredging, and overfishing. Loss of seagrass beds removes a large carbon sink, accelerating the beginning of hazardous climate change consequences on fish and fisheries in a recently known feedback loop.

CHAPTER 8

RECIRCULATING FISH FARMING

CONTENTS

8.1. Introduction 176

8.2. Ponds 177

8.3. Recirculating Systems 178

8.4. Hazards 180

8.5. Hardness 181

8.6. Water Flow 183

8.7. Diseases 186

8.1. INTRODUCTION

Over the last 30 years, recirculating aquaculture systems (RAS) (Figure 8.1) have evolved through university and commercial research and demonstration facilities, as well as ongoing improvement of each subsystem process. Aquaculture is still considered a viable alternative project in the United States, but it remains a substantial struggle to do so profitably. Indoor aquaculture presents an opportunity for food production in the United States, but it must be addressed. Seafood can only be as delicious as the water it lives in. Aquaculturists monitor the water quality to ensure that the seafood they produce is free of toxins. Consumers have shown a strong preference for farmed seafood, which is more consistent in quality and appearance and has a milder flavor than wild seafood. Depending on the quantity of organisms cultivated per volume of water and the water source and supply, aquaculture systems can be extensive, semi-intensive, or intensive. Pond culture is extensive, cage culture is semi-intensive but intense within the cage, and RAS systems are intensive. Because pond and cage systems are open-air, pollutants from the air or water are always a possibility. Because pond and cage systems have more challenging water quality control, the number of species that can be cultured effectively is limited. In the open air, the concepts of RAS for the water environment can be used, but you lose entire control of the environment. Outdoor ponds and net pen-based systems have a substantial disadvantage in terms of disease risk, which could lead to culture loss. Diseases in fish systems are spread through direct contact with infected organisms in the water (Espinal and Matulić, 2019).

Figure 8.1. Recirculating aquaculture systems.

Source: https://upload.wikimedia.org/wikipedia/commons/thumb/9/95/Recirculating_Aquaculture_System_7.jpg/800px-Recirculating_Aquaculture_System_7.jpg.

8.2. PONDS

Outdoor pond and net pen-based systems are additionally hampered by their inability to offer a consistent product due to problems with controlling growing cycles, resulting in supply peaks and troughs on the market. Problems with escapement are a big concern, especially with biotechnology modified species. Indoor RAS allows for controlled product growth rates and predictable harvesting schedules by raising fish in a controlled environment. RAS saves heat and water by reusing water after it has been reconditioned through biological filtration with biofilters. RAS enables effective economies, resulting in the highest productivity per unit area. RAS are environmentally friendly since they utilize 90–99% less water than traditional aquaculture systems, take up less than 1% of the land area, and provide environmentally safe waste handling. Others current commercial RAS utilize less water, have better densities, and similar feed conversions. RAS enables year-round production of predictable volumes of product as well as comprehensive environmental climate control. Aquaculture systems created by RAS are indefinitely scalable. Because waste streams may be controlled in environmentally friendly ways, there are no environmental restrictions on the size of the proposed fish farm. RAS provides extensive environmental control. This not only decreases the risks associated with outdoor aquaculture, but also allows for year-round species growth. In the domestic poultry sector, where chickens were brought indoors and the expense of environmental management was more than offset by greater growth rates, improved feed conversion, and more efficient labor utilization, a similar optimization was observed.

Even in Idaho, where government agencies permit limit fish farm effluents to monthly average suspended solids concentrations of 5 mg/L, serial-reuse systems may not need to capture large percentages of waste solids for their average discharge to meet concentration-based effluent limits on TSS or total suspended solids. Nutrient restrictions are also becoming a concern with discharge. However, because of the huge water volumes used in single-pass and serial-reuse production systems, removing nutrients from effluents is not feasible for farms like Clear Springs Trout Company. As a result, some treatment is employed to ensure that spikes in TSS are not discharged, even though overall waste capture efficiencies in serial-reuse systems may only be 25–50%. A commercial aquaculture business's success is in creating the ideal environment for rapid growth at the lowest possible cost of resources and money. The capacity to regulate the aquatic environment and crucial water quality parameters to improve fish health and growth rates is one of the major advantages of intensive recirculation systems. Despite the fact that the aquatic environment is a complex ecosystem with numerous water quality variables, it is fortunate that just a few of these variables are critical. Temperature, pH, and dissolved oxygen (DO), ammonia, nitrite, CO_2, alkalinity, and suspended particles concentrations are all significant characteristics. Each characteristic is significant in and of itself, but the totality and interdependence of all parameters determine the fish's health and growth rate. Each water quality metric interacts with and influences other parameters in a variety of ways, some of which are complex.

Not only is there less oxygen accessible to the fish in this situation, but the fish are also less able to use the oxygen that is available. The fishes' circulatory capacity to carry oxygen is affected by the high carbon dioxide level in the water, exacerbating the stress caused by low DO levels. The link between pH and ammonia toxicity is another outstanding example of the complicated interaction among water quality factors. At low pH, ammonia in the water is in the non-toxic ionized form. However, a one-unit increase in pH, from 6.5 to 7.5, raises the hazardous unionized ammonia concentration by a factor of 10. Simply increasing the alkalinity of a system with baking soda can cause the unionized ammonia to rise to hazardous levels. Some of the engineering principles that follow require an awareness of certain physical aspects of water. Water has various special qualities that allow life to exist on the planet. At the freezing point, the density drops dramatically, causing a big discontinuity. Ice has a lesser density than water and floats. Lakes and oceans would freeze from the bottom up if water did not exhibit this peculiar density pattern connected with temperature, making life as we know it impossible. The density of water is also increased by adding salt and other contaminants. Beyond the minimum exchange volume, the additional amount of fresh water required for any given system is directly proportional to the degree of reuse or recycling of the existing water. Apart from the obvious reduction in water demand and lower discharge water that must be treated, there are other advantages to reuse supply water as much as feasible.

8.3. RECIRCULATING SYSTEMS

Serial-reuse systems, partial-reuse systems, and fully recirculating systems are the three types of reuse systems. Fully recirculating systems reuse a higher percentage of total system volume than partial-reuse systems, and partial-reuse systems reuse a higher percentage of total system volume than fully recirculating systems (Summerfelt et al., 2004). The rate of depletion or accumulation of critical water quality metrics is affected by the amount of water reused. The higher the reuse rate, the more conditioning the reused water requires to return the water quality to target levels. The concentration of DO is usually the most important and first recognized limiting factor that determines the density of fish grown in a system. The concentrations of unionized ammonia and dissolved carbon dioxide levels are the next most critical components. These two variables are inextricably linked. This is owing to the direct impact of dissolved carbon dioxide on pH and the link between pH and ammonia-nitrogen toxicity. The pH rises as the dissolved carbon dioxide levels fall, increasing the toxicity of the total ammonia-nitrogen in the system. The selected reuse system must be able to maintain required levels of DO while also keeping dissolved carbon dioxide, unionized ammonia, and pH below their respective limitation values. The limiting water quality element in trout and salmonid raceway production systems is usually the DO concentration between racetrack sections, hence serial-reuse systems have been widely adopted. The water can be reused in the next portion of the raceway by adding oxygen until the accumulated ammonia levels become too high. Due to the requirements for more advanced oxygenation and monitoring systems, this basic design has substantially boosted raceway production, but at a greater system cost.

A high-quality water source with sufficient capacity to meet initial needs and anticipated future expansions is one of the most crucial prerequisites for a successful aquaculture plant site. As a result, while selecting an aquaculture site, the water supply must be properly evaluated and quantified in both quality and quantity. Aquaculture differs from other agricultural activities in that it necessitates a steady supply of high-quality water. Because most wells are only used on a regular basis, a standard well pump test is usually only done for a few minutes to determine the well's ability to deliver a specific amount of water in a specific amount of time. During normal aquaculture operations, however, the well is continuously pumped. As a result, the well's actual output may be substantially lower than the short duration test indicates. Aquaculture operations can use a variety of water sources, each with its own set of benefits and drawbacks. Because the goal of a RAS is to use as little water as possible, groundwater and municipal water supplies are the two most common water sources. Both have the needed quality, quantity, and reliability, thus the decision between them is primarily on availability and cost. Because of the higher dangers of contamination by pollutants, fish eggs, insect larvae, disease germs, and wide seasonal temperature changes, surface waters are rarely used. Groundwater sources have the advantage of maintaining a steady temperature throughout the year. Shallow groundwater sources, on the other hand, resemble the area's typical air temperature. The geology of the area surrounding the source has a direct impact on the chemistry of groundwater. Groundwater in limestone locations is rich in calcium and inorganic carbon. Groundwater in granite formation sites is soft and poor in dissolved minerals and inorganic carbon (Smedley and Kinniburgh, 2017).

8.4. HAZARDS

The high amounts of dissolved hazardous gasses such as hydrogen sulfide, methane, and carbon dioxide in groundwater, especially from deep wells, are drawbacks. Because of biological activities occurring in the aquifer recharge zone, most groundwater has little or no DO. Furthermore, groundwater is frequently supersaturated with nitrogen gas, argon, and carbon dioxide once pumped to the surface. To remove excess gasses and supply oxygen, the groundwater must first be conditioned by some type of aeration and degassing. High dissolved iron concentrations can potentially be an issue. A municipal water supply is another possible source of water. However, it's important to remember that these sources are designed and processed to protect human health, which means chemicals like chlorine and fluorine are added. These compounds must be neutralized either chemically or by filtering before being used in RAS. Chlorine is toxic to fish, and even low amounts of 0.02 mg/L are stressful (Leal, Neves, Santos, and Esteves, 2018).

The water quality requirements used in the design of culture systems are heavily influenced by the project's goals, species, and life stage. The most essential factor is the effect of water quality on growth. Fine quality and appearance, which is vital for live-market species; tissue quality, such as texture, taste, and odor; and finally, trace contaminants in the tissue, such as PCT, DDT, or dioxins. Water quality guidelines for specific species or systems take time and money to develop. Due to the ambiguity about their impact on production, several other water quality criteria are also being thoroughly examined. High amounts of fine particles and organic chemicals are unknown at this time; however, they have been linked to disease outbreaks. The leaching of surface-active chemicals from uneaten feed and feces, as well as the resultant decrease in surface tension, could be a significant design consideration. The corrosion of pipes and fittings, as well as vitamin premix, have the potential to build up dangerous quantities of some heavy metals, such as cadmium, copper, and zinc, in limited exchange systems. Heavy metal toxicity is highly influenced by water chemistry. Finally, information on the effects of pheromones, endocrine-disrupting chemicals (EDCs), and chemicals leached from paints, PVC components, and liners is scarce. These substances can have a significant negative impact on aquatic species by interfering with their endocrine systems, potentially resulting in lowered fertility and development rates. The ambient chemistry of several EDCs is complicated, and analysis is expensive.

The most significant and critical water quality characteristic is DO, which requires regular monitoring in intense production systems. Despite the fact that humans breathe air that contains 21% oxygen is very marginally soluble in water. As a result, aquatic animals must invest significantly more energy to take DO from water than land-based species must expend to obtain oxygen from the air. Furthermore, when salinity rises, oxygen solubility decreases. Oxygen concentration is affected by both barometric pressure and altitude. Because the response to low DO is not life-or-death, but rather a continuum of physiological impacts, defining essential DO values is problematic. Furthermore, exposure period, fish size and health, water temperature, carbon dioxide content, and other environmental factors all influence these effects. Warm water fish, on the other hand, feed best, grow fastest, and are the healthiest when DO levels are above roughly 5 mg/L. However, DO concentrations above this saturation threshold appear to provide little advantage to fish (Srithongouthai et al., 2006). When environmental DO concentrations are at the necessary concentrations, the gills can only transfer a certain amount of oxygen to the blood, and they are extremely close to or at their maximum transfer capabilities. Higher oxygen levels in the water have no effect on the amount of oxygen carried by the bloodstream.

8.5. HARDNESS

Hardness refers to the ability of water to precipitate soap; the harder the water, the more soap must be applied to a given volume of water to have the same cleaning effect. Hardness is measured in mg/L equivalent calcium carbonate and is described chemically as the total content of calcium (Ca^{2+}), magnesium (Mg^{2+}), iron, and manganese. Chemical titration is used to determine hardness in practice. Natural waters have total hardness ranging from less than 5 to over 1,000 mg/L $CaCO_3$. Soft (0–75 mg/L), moderately hard (75–150 mg/L), hard (150–300 mg/L), and very hard (> 300 mg/L) are the conventional classifications for water ("Hardness of Water | U.S. Geological Survey," n.d.). Because both are generally characterized in terms of mg/L $CaCO_3$, hardness, and alkalinity are frequently confused. In reality, if the alkalinity in the water comes from limestone, the hardness and alkalinity concentrations can be comparable, if not identical. In contrast, many groundwaters in coastal plain areas have high alkalinity and low hardness concentrations. Because of the relative low solubility of these minerals, aquifers in basalt and granite regions frequently have low total hardness and alkalinity. If this water has a low hardness and alkalinity, it must be hardened by adding dissolved calcium to sustain newly fertilized freshwater fish eggs and calcification of larval skeletal structures. The toxicity of dissolved metals is also reduced by calcium and magnesium. Total hardness recommendations range from 20 to 300 mg/L.

Fresh, brackish, and salty are all terms used to describe water. Each of these categories refers to the salinity of the water, and there are no apparent cutoff points between these classifications. The overall concentration of dissolved ions in water is known as salinity, and it is measured in parts per 1,000 (ppt) or grams of salt per kg of water. Calcium, sodium, potassium, bicarbonate, chloride, and sulfate are the principal sources of dissolved ions. Rainbow trout fingerlings, for example, are raised in freshwater, acclimated to saltwater, and then developed to marketable size in marine cages with salinities as high as 20 parts per 1,000. Most commercially important freshwater fish breed and grow successfully at salinities of at least 4–5 ppt. Fish regulate the intake of ions from the environment and limit ion loss to maintain the concentration of dissolved salts in their bodily fluids. This is referred to as osmoregulation. Because their bodily fluids are higher in ions than the surrounding water, freshwater fish, for example, accumulate water. When aquatic organisms are subjected to salinity levels outside of their optimal range, they must devote a significant amount of energy to osmoregulation at the expense of other processes, such as development. The animals cannot maintain homeostasis if salinity deviates too far from optimal.

Uneaten feed fines, fish feces, algae, and biofilm cell mass sloughed from biological filters are all sources of waste solids in aquaculture systems. According to studies, for every 1 kg of feed provided, fish create 0.3 to 0.4 kg TSS. In a recirculating system, waste solids have an impact on the efficiency of all other unit processes. They are a substantial source of carbonaceous oxygen demand and nutritional input into the water, and they can harm fish gills and house pathogens in recirculating systems, compromising fish health. TSS levels of 80 mg/L are acceptable for tilapia if all other water quality indicators are within acceptable limits. One of the most critical operations in aquaculture systems is solids removal. Solids should be removed as soon as feasible from the fish culture tank, with as little turbulence and mechanical shearing as possible. Solids can be divided into three types: settleable, suspended, and fine or dissolved solids. The time it takes for settleable and suspended solids to settle to the bottom of the cone is the only difference. Although the Winkler Method and a simple titration can be used to measure DO, today's DO meters allow for rapid and accurate analysis over a range of zero to supersaturation. Polarographic and galvanic electrodes are both extensively employed. Both meters have an electrode that provides a signal proportional to the concentration of oxygen in the water, as well as hardware that converts the signal reading to a visual display or recording device. A gold electrode and a silver-silver oxide reference electrode make up a standard polarographic DO electrode or probe. The electrodes are immersed in 4 M KCl and separated from the sample by a Teflon, polyethylene, or fluorocarbon membrane. When the probe is charged, molecular oxygen flowing across the membrane reacts with the cathode, i.e., gold ring, causing current to flow to the anode electrode, i.e., silver. The electrode in a galvanic system produces a low voltage proportionate to the DO content. Temperature, air pressure, and salinity adjustment are required in both systems, which is normally performed using a combination of calibration and instrumentation gear and software. There are currently a variety of DO meters available at various pricing points. The most recent DO measurement technology is based on dynamic fluorescence quenching of a luminescent dye. A luminous substance has been applied to the sensor in the cap. The luminous chemical on the surface of the sensor cap is illuminated by blue light from an LED. The luminous molecule becomes agitated almost instantaneously, and then releases red light when it relaxes. A photodiode detects the red light, and the time it takes for the chemical to return to its relaxed condition is calculated. The lower the oxygen concentration, the less red light is emitted by the sensor, and the faster the luminous material returns to its relaxed state.

8.6. WATER FLOW

Water flow is the mechanism that transports oxygen into a fish culture vessel while also removing waste products formed within. The main elements impacting water quality and fish production, such as oxygen, ammonia, carbon dioxide, and suspended particles, should be carefully balanced in the design of a RAS. To determine the thresholds for each of these parameters, the value of each must be calculated separately. The system must then be run at the lowest flow rate possible while maintaining a particular parameter at its design value, such as ammonia, after the relevant calculations have been completed. Obviously, the lowest flow rate required to maintain one parameter may be insufficient to maintain another. Any variable affecting water quality can be studied using the same mass balance approach. The only thing aquacultural engineers can agree on is that raising fish requires a body of water. The best way to control this body of water is still up for dispute. The most important consideration in all tank culture design is to enhance the tank system's ability to self-clean. The failure of RAS is almost always due to their incapacity to self-clean. Recirculating systems that do not effectively and quickly remove fish feces, uneaten feed, and other solids from the culture water will never produce fish economically, with the exception of the occasional heterotrophic or ODAS (organic detrital algae soup) based systems used for tilapia or shrimp. All other individual components of the system will fail to work efficiently if solids removal is poor. As a result, the first question to examine is whether the culture tank can properly remove solids. The capacity to successfully manage the fish within the culture tank is the next question to be addressed.

Many people assume that when a RAS is installed inside a structure, the heating/cooling bills will be high. The idea is that if water exchange rates are high, water heating expenses will be high as well. The majority of these assumptions are unduly pessimistic. Water exchange in a RAS can be kept to less than 20% of system volume per day, and frequently less than 10%. Water heating expenses will be a modest percentage of total production costs under these conditions. Water heating losses will, of course, be proportionate to the water exchange rate. Many RAS designers make the error of overlooking the ventilation requirements of the building in which the RAS is housed. The same way a ventilation balance is done for chickens or cows to ensure an appropriate air quality environment, ventilation requirements for controlling environmental conditions of humidity, temperature, and CO_2 should be tackled. Instead of warm-blooded animals, we have fish for RAS. The regulating environmental conditions in a RAS are normally humidity and then CO_2. Selecting a place will also establish a political district in which you will be required to comply with all applicable laws and regulations. So, the first step is to see if we can comply with the laws, and if we do, whether you still want to build a farm here. Make certain you are familiar with all of the rules. However, the system's supporting components, which are just as vital, if not more so, for commercial success, are rarely emphasized. All of the other pieces of a recirculating system that are required for its profitable functioning are referred to as support components. The success of a recirculating system is typically determined by how well these support systems are built, integrated into the operation, and managed. One common feature is that they are frequently overlooked when assessing the costs of constructing a cultural system. As a result, support systems are frequently implemented late in the construction process or as a last-minute addition following a disaster.

The electrical power supply, which is necessary to operate pumps, aeration systems, instrumentation systems, and a variety of other tasks in a recirculating system, is one of the key systems that requires backup. Electric supply failure can have disastrous consequences in minutes, especially in heavily loaded systems. A generator that runs on gasoline, diesel, natural or propane gas can offer backup electricity. For any commercial system, backup generators are necessary. Several commercial manufacturers can provide systems for a wide range of power requirements. Automatic systems or cut-over switches are costly, but they are required when workers are absent at particular times of the day or are unable to get to the plant quickly enough to perform manual emergency procedures. If continuous manual coverage is not possible, an automatic transfer switch will cost around 35% of the overall cost of the backup generator system, but it is generally well worth the expenditure. If you're going to use an auto-transfer method, make sure you have protocols in place to protect the fish if it fails. The alarm sensors will occasionally detect that power has been restored when it has not. Three-phase systems are particularly troublesome, and they fail in the transfer process more frequently than single-phase systems. Remember that an auto-transfer isn't as dependable as human coverage and a manual transfer switch. Generator maintenance is crucial. If power failures are rare, the backup generator may go unused for extended periods of time. Smaller generators should be started on a regular basis and kept in good working order at all times. This provides fire protection, inhibits heat buildup around the generator set, and reduces noise levels.

Every recirculating aquaculture production system, however, will necessitate at least a small amount of laboratory space. In any aquaculture operation, water quality monitoring and control is a standard procedure. The quantity of laboratory space and equipment required for this job will vary depending on the sample methodology, facility size, and sampling frequency. Commercial water quality test kits from a variety of vendors are frequently sufficient for smaller fish farms. They use chemicals that have already been packaged, indicator strips, or color comparison procedures. They are simple to use, cost little per analysis, and provide enough precision for small production plants with modest stocking densities. Furthermore, they take up very minimal laboratory space. As the number of samples and the associated production risks grow, the sophistication of the analysis must also grow. Temperature, pH, DO, conductivity, and oxygen reduction potential are just a few of the significant water quality characteristics that can be monitored using electrode sensors that produce an electrical output signal proportional to the water quality parameter being measured.

Portable meters do necessitate the physical presence of someone at each sampling location to manually measure and record the appropriate water quality parameter at whatever frequency is requested. In comparison to the equipment expenses for specialized systems, labor costs for manual monitoring might be expensive for large systems but low for small systems. Dedicated systems can be utilized to monitor and control essential water quality factors such as temperature, pH, and DO in intensively stocked systems. There are commercial systems that will monitor chosen parameters and alert you if they depart from the predetermined minimum and maximum values. Real-time water quality data is made available to the system manager when it is combined with the requisite computer gear and software. This gives managers precise data as well as previous data for evaluating and analyzing system performance. A higher-quality laboratory will be required as sample techniques become more advanced. For example, various commercially available laboratory spectrophotometers for water quality analysis are already available. Using pre-packaged chemicals and basic laboratory techniques, these dedicated pieces of equipment can analyze a wide range of key water quality indices. The laboratory should now have a sink, refrigerator, and computer, as well as enough space for equipment storage and personal work areas.

8.7. DISEASES

Diseases of fish purchased from outside the system cause many of the disease outbreaks in recirculating systems. Incoming fish should be quarantined for one to several weeks before being introduced into the recirculating system to reduce disease transmission. Any disease concerns can be handled during this time without polluting the recirculating production system. Quarantine areas require room and must be situated such that arriving fish do not come into touch with the growth systems. Such migration must be impossible due to permanent physical impediments. The quarantine room should be located in a separate building from the main production systems wherever possible. The size of the quarantine area is determined by the facility's management practices. The quarantine space must be large enough to accommodate all of the fish entering during a single quarantine period, whether that period is one week or several weeks. Water should be sufficient for the number of fish in the quarantine area, and waste should be discharged separately. The most appropriate procedure is determined by the waste's properties, concentrations, and form. Once the waste has been collected from the culture water, it must be disposed of or recycled. Solids from fish culture systems are high in nitrogen and phosphorus, both of which are important nutrients in fertilizers. As a result, one method of disposal is to employ the solids as fertilizers by spreading them on agricultural land. Both solid and liquid fish wastes can be composted for later use as a soil additive, mulch, or fertilizer. Holding facilities may be required to keep wastes contained between disposal cycles. These facilities must be able to confine waste in an appropriate manner while also preventing the development of odors and other harmful substances. Almost all aquaculture facilities will have some mortality, however the number of dead fish will be minimal in normal operation. However, if a large number of fish die as a result of oxygen deprivation, disease, or other factors, the number of dead fish that must be disposed of can quickly mount. These fish must be disposed of in an environmentally friendly manner, and they must not emit offensive odors or constitute a health or sanitation danger to humans or other fish. The volume of dead fish, land usage near the facility, groundwater table depth, and other considerations will all influence acceptable disposal strategies. Lack of planning for the disposal of dead fish in the case of a mass mortality can be awkward, and may result in a legal problem. Processing waste must be disposed of or utilized if fish is processed on-site. Composting, anaerobic digestion, burying, and transporting to a landfill are some of the alternatives for huge amounts of fish waste. The strategy is to plan ahead of time.

When grading fingerlings, the box grader is frequently used. It consists of a floating box with a set of grader bars that may be adjusted or replaced. In most circumstances, trial and error will identify the right bar width. Grader size can be connected with condition parameters calculated from weight and length samples after a few gradings, allowing for more precise grading planning. Records of fish size and grader bar width should be kept in the production notebook so that benchmarks connecting fish size, weight, or condition factor to grader width can be established. The grader floats in the tank with the fish to be graded on top of it. The smaller fish pass through, while the caught larger fish are dumped into a separate tank. The grading procedure can be made easier by gently raising and lowering the grader in the water. Three tanks are best for grading fish into two size groups, although two tanks may suffice. The source tank is progressively drained as the other two tanks are filled with little or large fish when three tanks are employed. Fish are packed to one side of the tank using a divider, such as a hinged screen, when two tanks are employed. After that, a portion of the fish is netted and placed in a grader that floats on the other side of the wall. Fish transportation is essential for fish survival. You must maintain your fish alive throughout shipping in order to run a successful business. When employing RAS technology, this often entails learning how to get healthy fish into the system and then send them to market alive. It's pointless to invest time and effort into developing and operating an efficient system if you don't also understand how to transfer animals safely. When live fish are transported, they are frequently neglected during transportation, resulting in stress, disease, and long-term mortality. When it comes to live hauling, the first thing to remember is that the species you're transporting will have an impact on how you move them. For successful transport, several species require varying water temperatures and loading rates. Two components of the business entail live shipment. They are fingerling transit from the supplier to the culture facility and product delivery to live markets. Fingerlings must be carried from the source to the grow-out facility in the first instance. To reduce stress-related damage and mortality, they must arrive in good condition. Dealing with reliable vendors who have a track record of providing high-quality fingerlings is critical (Portz, Woodley, and Cech, 2006).

When transferring fish outside of the facility, there are a number of options for reducing stress. Reduced water temperature, supplemental oxygen to keep concentrations at or above saturation, 0.5% salt, putting the fish off feed 24 to 48 hours before transport, changing the water between several journeys, and administering anesthetics like MS-222 are all strategies to alleviate stress during transport. Note that if MS-222 is used, the fish should be taken off feed for at least 24 hours before being transported, as the fish are more likely to defecate and pollute the water after exposure (Harmon, 2009). If you're hauling fish that will be sold for human food right away, you'll need to check with your local state authorities about withdrawal periods. MS-222 may cause stress in the fish upon its first exposure to the esthetic.

Because much of the stress reaction is induced by crowding, netting, and transport, immersing them in water with esthetic simply attempts to calm agitated fish. The material holding fish should be resistant to degradation caused by ultraviolet (UV) radiation from the sun, which can make some materials brittle and break within a year or two. Foam should be used as an intermediate insulation layer in good hauling tanks to assist maintain the temperature within the hauling unit. While well-built devices will require little maintenance, accidents do happen. When it happens, fiberglass is easier to repair without specialist equipment or skills than aluminum, which necessitates the use of sophisticated welding procedures. Non-toxic materials are essential not only because we do not want fish to die, but also because we are working with food goods in most circumstances, which should always be a top priority in our industry. Cleaning and sanitizing hauling equipment is critical and should be considered a routine element of maintenance.

Other shapes, on the other hand, have been employed in the past and may be more efficient to deal with. Circular tanks have been employed by some companies. Other fish hauling units have elliptical bottoms that are partially rounded. This aids circulation by mixing the water and preventing fish injury due to the lack of sharp corners. The elliptical portions on the sides can be filled with expanded foam to further insulate the tank from heat, and the form tends to concentrate weight over the center of gravity of the hauling vehicle, which can help to stabilize the ride. These tanks are also simple to unload. The difficulty of these units is that they lose volume and area. A circular tank, for example, may lose nearly 20% of the area that rectangular or square tanks intended to fit in the same space could cover. This could be a crucial issue if maximal fish capacity is important. Before making a final decision, the fish farmer should consider all options and research them thoroughly. Fish may be loaded and unloaded more rapidly with proper tank structure, resulting in lower stress levels for the fish. Large hatches at the top of the hauling unit should be hinged and allow enough space for placing fish into the hauling tanks rapidly. When the vehicle is moving, these hatches should have dogs or clamps to keep the doors shut. Watertight seals are routinely added to ensure that no water is lost while the vehicle is in motion. The majority of hauling tanks have many chambers, which can be built longitudinally or crosswise within the tank. During vehicle operation, these chambers prevent the free surface effect of enormous quantities of water traveling from side to side (front to back). If not checked, this motion has the potential to make the vehicle unstable. This is why the smaller sections are designed to reduce movement. In addition, the free surface effect in the tank will be reduced as the water level rises.

Unloading or dump gates should be installed on the tank to allow it to be emptied rapidly. Some tanks have sloping bottoms to funnel the fish down and out while unloading, although this structure consumes some of the tank's volume and may not be efficient when maximum capacity is needed. The drain should be flush with the tank's bottom to allow all of the water and fish to escape. If the drain has a lip, it will be difficult to completely empty the unit, and fish will most likely remain in the tank, making disinfection harder. Even if a little additional floor drain is created and only used for water drainage at the end of the operation, make sure the capacity to entirely drain a tank compartment exists. Before transporting fish within or between states, the operator must ensure that all necessary permissions have been secured for the species being transported. This may entail contacting natural resource management organizations in the states through which the transit will take place to ensure proper labeling and paperwork. Furthermore, several states have requirements for markings on fish carrying trucks.

Pathogen entry through a facility's water supply is a common route of infection, and it raises the danger of infectious disease outbreaks in aquaculture production systems. Groundwater should be used for the facility wherever possible. Resident fish, other aquatic animals, or aquatic invertebrates that could be infection carriers are rarely seen in wells and springs. Input water should be disinfected using UV radiation or ozonation if a pathogen-free water source is at danger of contamination or unavailable. Before using well or spring water for fish cultivation, it may be necessary to remove carbon dioxide and/or nitrogen gas, as well as supply oxygen. Before purchasing the property, the well should be drilled and tested. If the site does not have sufficient well water, a different location should be chosen. If a sufficient supply is available, well water is also the best option for larger operations. Because surface waters contain fish diseases, they should only be used as a last resort and only after adequate sterilization. Fish, birds, raccoons, salamanders, frogs, and snakes, among other species that can transport fish viruses, should be kept away from spring water.

CHAPTER 9

REPRODUCTIVE AQUACULTURE

CONTENTS

9.1. Introduction 192
9.2. Ovary 193
9.3. Reproduction 195
9.4. Spawning 198
9.5. Sodium Ions 201
9.6. Sperm Motility 202
9.7. EGG Quality Management 208

9.1. INTRODUCTION

The reproductive cycle is a sequence of events that begin with immature germ cells and end with the creation of mature gametes, with the goal of getting a fertilized egg after sperm insemination (Figure 9.1). Gametogenesis is the process of gamete growth and differentiation that leads to the creation of the female oocyte or male spermatozoon. Female and male gametes share the same start in a population of embryonic primordial germ cells or PGC that move to the germinal epithelium throughout embryonic development to create gonads. The PGC divide into gonidia during mitotic divisions, which differentiate into oogonia or spermatogonia depending on the individual's sex. Gonidia undergo meiosis and form oocytes or spermatocytes with the last mitotic cycle. The reproductive cycle in males and females has two key phases: gonadal growth and development and maturity, which ends in ovulation/spermiation and spawning. Spawning is a highly synchronized event that leads to fertilization of the egg and embryo development. The success of reproduction is determined by the successful completion of each step of the reproductive cycle, which results in the creation of high-quality gametes ("Methods in Reproductive Aquaculture," n.d.).

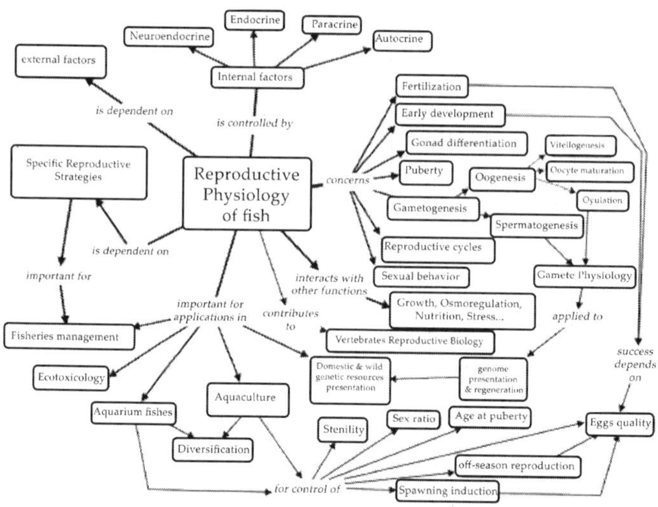

Figure 9.1. Reproductive aquaculture.

Source: https://images.thefishsite.com/fish/legacy/files/articles/old/08-10-31Fish1.gif.

9.2. OVARY

The ovary is a bilateral elongated structure located in the abdominal cavity of female fish. The mesovarium surrounds the ovarian lobules, which project posteriorly through a pair of oviducts to the genital papilla, which opens to the outside. The ovigerous lamellae are folds of the germinal epithelium that protrude transversally into the ovarian lumen, compartmentalizing the ovaries. Oocytes go through several stages of gametogenesis in these lamellae until they mature into ova, i.e., eggs that are discharged into the ovarian cavity or abdominal cavity at ovulation and then released into the external environment during spawning. Before spawning, an ovulated ova may stay in the ovarian or abdominal cavity for some time. They keep their fertilizing capacity for a while, but if they aren't spawned, the ova become over-ripe through a degeneration process. Stripping should be done before over-ripening develops in farmed fish whose reproduction is based on manual egg stripping and artificial insemination (Figure 9.2).

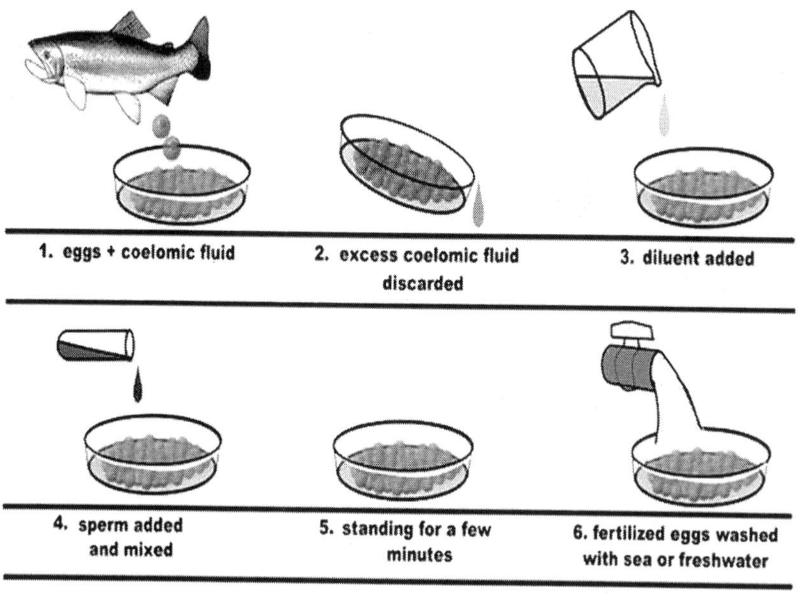

Figure 9.2. Artificial insemination.

Source: https://www.researchgate.net/profile/Jc-Gwo-2/publication/285635014/figure/fig5/AS:869964416294916@1584427088806/General-procedure-for-artifical-insemination-diluent-method-After-Leung-and-Jamieson_W640.jpg.

An egg is surrounded by two layers of follicular cells in the ovary's germinal unit. These follicular cells encircle the germ cell and provide structural and functional support to the developing oocyte, mediating external molecule internalization and producing hormones and components required for oocyte differentiation, development, and survival. An inner monolayer of granulosa cells and an outer monolayer of theca cells surrounds each oocyte. A thin basal membrane exists between the two follicular layers, separating them. The oocyte is also surrounded by a thick acellular envelope, to which the granulosa cells are linked. During gametogenesis, the zona radiata thickens and compacts, eventually forming the egg chorion or egg shell. The process of vitellogenesis, the synthesis of vitellogenin or VTG, the precursor of the egg's vitellin reserves, is unique to the female reproductive cycle of fish. VTG is a lipo phospho glycoprotein produced in the liver in response to estradiol stimulation or E2. It is released into the circulation and progressively integrated into expanding oocytes by receptor mediated endocytosis, where VTG is proteolytically broken into smaller components like phosvitin, lipovitellin, giving rise to the egg's vitellin reserves, yolk, or vitellus. The correct accumulation and processing of VTG is critical for the quality of the egg and subsequent survival of the born larvae, as this is the larva's only source of nourishment until external feeding begins several days after fertilization. The VTG is a female-specific protein, and its blood levels are closely linked to the start and progression of the gametogenic cycle.

The dissolution of the nuclear membrane, known as GV break down, marks the end of oocyte maturation. The ionic composition of the cytoplasm is altered by the alteration of the lipid and yolk inclusions, resulting in a significant increase in water incorporation into the oocyte due to increased osmotic pressure. This extreme hydration is especially important in fish that produce pelagic eggs, since it induces a rapid rise in oocyte volume of 2–3 times. The follicular wall ruptures after hydration, allowing the oocyte to be ovulated into the ovarian or abdominal cavity and released into the water during spawning. Post-ovulatory follicles or POF can be seen in the ovary during the spawning season. Oocytes that are apoptotic can also be observed. These are oocytes that were unable to complete the vitellogenesis or FOM phase due to a hormonal malfunction in the reproductive process. The vitelline envelope is fragmented when the oocytes die, and hypertrophied follicular cells infiltrate the ooplasm for phagocytosis. Under the microscope, follicular atresia appears as a compact, well-vascularized structure. The quantity of atretic oocytes increases during the pre-spawning and spawning seasons, indicating the female's reproductive success. It is found in all species and is related to the reproductive environment's suitability. Follicular atresia can occur at any stage of oogenesis and controls the number of oocytes that go through the reproductive process, affecting the species' fertility. In aquaculture, determining the stage of gonadal development in female breeders is a vital tool. Biopsies of oocyte samples can be examined to detect this. A catheter is inserted via the gonoduct and gentle aspiration of intra ovarian oocytes is conducted in sedated females. The oocytes are examined under a binocular microscope and classified according to their size, GV position, degree of yolk granule coalescence, and other factors; these classifications provide a relative indicator of the females' gonad development stage.

As they all stem from a single initial cell, this cluster of daughter cells is created by a clone of spermatogonia. A wall of somatic Sertoli cells surrounds each cyst, keeping different clones isolated from one another. During this phase of mitotic proliferation, spermatogonia go through a weak division rate and subsequently a rapid division rate. The final mitotic division of spermatogonia produces primary spermatocytes, which enter the meiosis process. In phase two, primary spermatocytes go through their initial meiotic division, which involves DNA duplication and recombination of genetic material, resulting in secondary spermatocyte creation. They proceed to the second meiotic division quickly and without duplication of DNA, resulting in the creation of haploid germ cells, the spermatids. The haploid spermatids enter the spermiogenesis process, when they develop into flagellated spermatozoa. This process does not entail cellular proliferation, but rather cell transformation, which includes a significant reduction in size due to nucleus condensation and cytoplasmic extrusion into the surrounding Sertoli cells. At this time, mitochondrial concentration, the creation of a mid-piece, and the production of the flagellum all occur. Because fertilization is accomplished through the micropyle of the fish egg, all but a few species of fish lack the creation of the acrosome found in all other vertebrates.

9.3. REPRODUCTION

The goal of reproduction is to produce viable progeny. Food availability and environmental conditions have long been considered as factors that influence survival and, thus, how a species evolves through natural selection. When a fish matures, it is determined by the availability of food and its ability to retain energy. According to some theories, fish can determine what size and/

or age parameters are best for completing maturation during the first and future maturation episodes by using a genetically determined bio-chemical threshold. The timing of reproduction is determined by the availability of food for young and thus offspring survival. Food availability varies seasonally in the tropics, both at higher latitudes and at lower latitudes. Reproduction is scheduled to ensure that important feeding intervals for offspring survival, particularly larvae and juveniles, coincide with seasonal periods of abundant food. This is frequent in species that are in the high latitudes, and it is often in relation to oscillations in nutrient levels induced by changes in ocean currents, temperature cycles, or weather cycles such as rainy seasons for species in the lower latitudes. As a result, maturation is a complicated process that must be precisely planned in order for spawning or important feeding phases to coincide with seasonal highs in food supply months or even years after maturation began. The quality of pre-frozen sperm is crucial for successful cryopreservation (Figure 9.3).

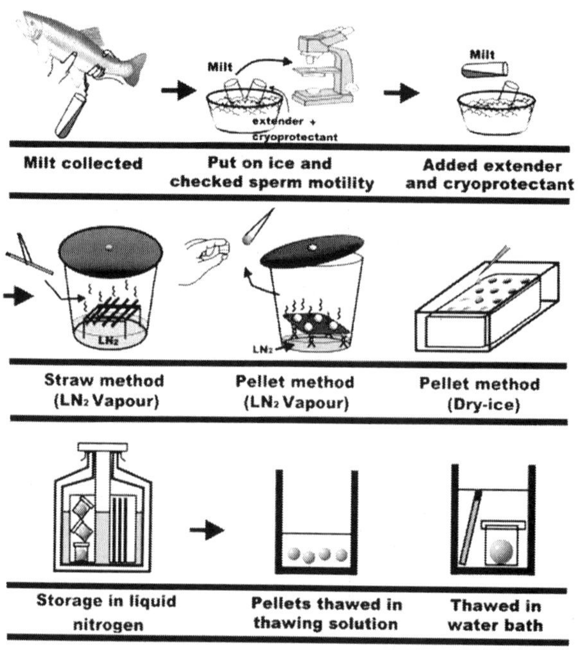

Figure 9.3. Cryopreservation.

Source: https://www.researchgate.net/profile/Jin-Chywan-Gwo/publication/267193203/figure/fig7/AS:667089622810627@1536057968193/Cryopreservation-procedures-for-fish-sperm-pellet-straw-and-vial-methods-Adapted-from_W640.jpg.

The structure and cryo-resistance of spermatozoa, the physicochemical composition of seminal plasma, and the quality and volume of semen differ significantly between species, strains, and even individuals within a single stock (Agarwal, 2011). Furthermore, spermatozoa fitness differs depending on the animal's health, food, storage period, storage conditions, collection techniques, and season. As a result, the quality of the pre-frozen sperm from the chosen male should be scrutinized. Incision is a highly effective method for removing testes and extracting sperm. The bloodstock should be kept stress-free to optimize the quality of the collected semen. This includes collecting sperm during the spawning season, selecting sexually mature and healthy bloodstock, anesthetizing animals during handling, and limiting any human injury to the animals. It's best to keep sperm undisturbed after collection to avoid depletion of their energy stores before freezing, and to keep semen from combining with contaminants. Semen samples should be maintained at 4–5°C and frozen as soon as feasible after collection under aerobic circumstances. Once removed from the fish, sperm deteriorates quickly.

To extract sperm from finfish, the fish must be removed from the water and any traces of mucus and water must be cleaned off its body, with the urogenital opening area or genital papilla kept completely clean and dry. The abdomen of the fish should be gently stroked to eliminate urine and fecal contamination before collecting semen. Male broods with gushing sperm should be chosen first. It's critical to avoid contamination with water, slime, scales, blood, bile, or peritoneal fluid when collecting sperm via gentle abdominal stripping. Contamination can be avoided by collecting sperm with a catheter.

Semen can be retrieved from certain fishes, such as the ocean pout or *Macrozoarces americanus* (Figure 9.4), by inserting a polyethylene tube straight into the sperm duct through the urogenital opening. When the sperm sample is contaminated, centrifugation may be required. Tilapia sperm is centrifuged for 5 minutes at 5,000 rpm, then the supernatant is decanted. The sperm motility is unaffected by centrifugation at 1,000 g (Pérez-Atehortúa et al., 2021). Finally, keep each fish sperm on crushed ice until motility and quality are determined, then pool if necessary. Semen should not be exposed to direct sunlight. During the spawning season, purchase mature, gravid adult broodstocks from hatcheries. Because it is difficult to establish the sex of oysters and sea urchins by observing external appearance, determine the sex of abalone or *Haliotis diversicolor* by external examination on the gonad index. To sample a small bit of the gonad, drill a hole in the Pacific oyster or Crassostrea gigas shell or remove the top valve, or inject a syringe through the soft part around the mouth of the sea urchin. Examine a smear of gonad material under a light microscope at 200 x magnification to determine the sex and sexual maturity of individual animals. To avoid unexpected spontaneous spawning and possible gamete cross-contamination, keep animals in separate aquaria or tanks with aerated seawater at constant and natural water temperature, salinity, and photoperiod. Before harvesting gametes, scrub and rinse the exterior surface of all adult mollusks like oysters and small abalone in fresh seawater. Remove the spines from the sea urchin shell, then wash and dry the denuded shell with seawater. This reduces the chance of infection from ciliates and other protozoa found on the shell, as well as contamination from water, feces, and mucus.

Figure 9.4. Macrozoarces americanus.

Source: https://upload.wikimedia.org/wikipedia/commons/thumb/b/ba/WHSA_Ocean_Pout.jpg/800px-WHSA_Ocean_Pout.jpg.

9.4. SPAWNING

Dissection or induced spawning can be used to extract sperm. After removing the oyster shell, scrape the gonad and place the tip of a pipette or the needle of a syringe into the incision to remove the gametes. Semen can be extracted from larger oysters by inserting a syringe into the gonad through a hole drilled in the shell, which will not harm the animals. Using tweezers, grasp the slightly protruding spermatophores and extract them from the vaginal pore's terminal ampoules. Avoid causing harm to animals or interfering with their ability to produce fresh spermatophores. Sperm quality and quantity are unaffected by successive spermatophore removal. After mating, sperm can be collected from the spermatophore or sperm mass stored inside the female shrimp's seminal receptacle. Penaeoidean female shrimps either store sperm on the exterior of the *thelycum* or within the *thelycum's* exoskeletal invagination. Open *thelycum penaeoideans* mate within a few hours of spawning, and their sperm do not require post-testicular alteration to become fertile after transfer to the female. After a broadcasted release from males into the surrounding waters, their sperm is capable of fertilization. The spermatophore or sperm mass received at mating is retained for extended durations in closed *thelycum penaeoideans*, and sperm inside the spermatophore or sperm mass undergo capacitation following translocation to the female. Close *thelycum* shrimps and anesthetize them with ice before cutting through their thoracoabdominal segment right below the fifth pair of walking legs, where the *thelycum* is located. From the dissected thelyca, separate the whitish gelatinous sperm clumps. Press a spermatophore's posterior portion, rupture it, and release the sticky sperm mass that forms a droplet. To release sperm, mechanically homogenize the spermatophore with a tissue grinder or sperm bulk in seawater. Several rounds of resuspension and centrifugation will not harm sperm. Because mature crustacean sperm are nonmotile, numerous approaches for determining male reproductive potential and sperm cell viability in penaeid shrimp were compared.

The sole definitive criterion for assessing sperm viability is the induction of the acrosome response. The acrosome reaction of sperm is induced by egg water, a jelly precursor produced by eggs during spawning in seawater. Egg water can thus be used to investigate sperm maturation in both open and closed *thelycum* shrimps. Aquacultured species range from echinoderms to teleosts, but they all have one thing in common: they spend their entire lives spawning in water. The technique of fertilization and the kind of eggs greatly conditioned spermatozoa features and sperm physiology in sperm from these groups. Numerous strategies have been evolved to facilitate sperm-egg contact, ranging from the simultaneous spawning of many individuals in the same location to the discharge of milt over the spawn, which had previously been liberated by the female in a small nest. Internal fertilizers, or spermatozoa that enter the female body, are another category. Internal fertilization occurs in crustacea decapoda, such as shrimp, crabs, and lobsters: immotile spermatozoa are packed into a spermatophoric container, the spermatheca, which is transferred to the female's seminal receptacle, or *thelycum*, where sperm cells undergo physiological changes required for further fertilization, as previously demonstrated in shrimp. Males transfer sperm packed in spermatophores to females, which store the sperm in the receptaculum seminis of the genital tract for as long as several months after insemination. Some teleosts, including live-bearing fish such as swordtails and platy fish of the genus Xiphophorus, which produce ornamental fish as well as a model for developmental research, are also internal fertilizers and use a similar strategy. The structure of eggs from aquacultured species is likewise quite different. Almost all teleosts contain enormous eggs enclosed by a hard and impermeable envelope called the chorion. Depending on the species, the chorion is punctured by one or more channels, allowing spermatozoa to penetrate the plasma membrane and fertilize the eggs. Given the relative sizes of sperm and eggs, as well as the limited number of micropyles, many spermatozoa will never reach them and will therefore have no opportunity of fertilizing the egg. Because spermatozoa enter through a channel, direct contact between egg and sperm plasma membranes occurs, and the spermatozoa heads do not require particular enzymes to lyse the egg envelopes, as do those found in mammalian spermatozoa's acrosome.

When comparing the anatomy of spermatozoa, the head is the cellular compartment with the most variation in appearance. Because the existence of a small number of spherical mitochondria is typical in external fertilizers, such as sea urchins, bivalves, and teleosts, the mid-piece is commonly underdeveloped. Some teleosts in the genus *Ictalurus* have biflagellate spermatozoa, and some crustaceans have aflagellate spermatozoa, but uniflagellate gametes are the most prevalent. The testes generate sperm. During their journey through the efferent duct system, spermatic ducts produce and secrete enzymes, monosaccharide, lipids, and proteins into the seminal fluid, regulate plasma ionic composition, and exhibit phagocytic activity, allowing spermatozoa to mature and remain viable until spawning. Milt release is triggered by a variety of triggers and, depending on the species, can be restricted to a single breeding season or occur all year. Aquasperm cells are non-motile in the seminal plasma, and motility is induced when they are released into the water or, in the case of some ascidians, even later, when they come into touch with egg secretions. The activation of motility is a complicated process that requires more explanation. Most species' motility lasts only a few minutes, however spermatozoa from teleosts like herring and chondrosteans like sturgeon keep their motility and fertilization ability for several hours following activation. Aquasperm are relatively simple cells with a low energy production efficiency that are released into a media that does not allow them to survive for lengthy periods of time. In such circumstances, long-term movement mechanisms are unnecessary, and fertilization success is dependent on the discharge of a large number of spermatozoa extremely close to the egg, as well as possible processes that facilitate their interaction. After sperm discharge into the vaginal tract, motility begins in viviparous and ovoviviparous species. Semen from viviparous teleosts has uncommon characteristics, such as well-developed mitochondria and glycolytic activity comparable to mammalian sperm, allowing for long-term motility and survival in the female reproductive system, as well as being more mammalian-like in shape and behavior.

9.5. SODIUM IONS

Sodium ions are known to play a secondary role in the activation and regulation of sperm motility in fish and invertebrates. Indeed, there is little literature on this topic, and the role of Na^+ in the motility initiation and regulation process of spermatozoa was just recently revealed. Spermatozoa are immotile when expelled from the testes and when they come into contact with seawater. However, spermatozoa acquire motility when they come into touch with eggs. Under the regulation of ion concentration inside and outside the cell, this regulatory mechanism would allow spermatozoa to switch from forward to backward functioning channels. Because this species' motility may last longer than 60 minutes, it appears that the cells could manage the increase in calcium within by acting in reverse mode, allowing calcium ions to be balanced by entering sodium ions. This would allow for prolonged motility as well as the maintenance of sperm in a quiescent state in contact with seawater prior to contact with the specific egg factor. In some species, changes in pH after sperm release activate motility. Sea urchin spermatozoa are unable to swim in the gonads due to high CO_2 tension in the sperm, which keeps the pH of the gonads acidic in comparison to sea water. Spawning causes a rise in pH and activation of dinein. Then, ADP synthesis triggers a 50-fold increase in mitochondrial respiration and motility. Another recent work in *Asterina pectinifera* starfish found that increasing intracellular pH caused phosphorylation of the axonemal proteins necessary for flagellar activation. It's unclear how or whether pH from external solutions influences fish motility initiation. During motility activation, an elevation in intracellular pH has been observed in some teleosts, especially marine species. The pH range for motility activation in almost all species is quite wide, ranging from acid (pH = 5) to alkaline (pH = 10). As a result, it's possible that external solution pH regulates motility, however certain species have reported interference. This is the case in sea urchins and other species with acrosomal reactions and chemotactic behavior, in which internal pH rises in response to acrosomal reactions or chemoattractant exposure, but it is not a requirement for calcium entry into the cell in some of them, implying that the regulatory effect of pH on sperm activation has yet to be demonstrated. In numerous species, temperature from the external media alters the motility pattern and duration. Low temperatures (LTs) cause motility to last longer, with a decrease in cell velocity and an increase in the frequency of flagellar beats.

9.6. SPERM MOTILITY

Sperm motility can occasionally be boosted by secretions from the eggs or female reproductive organs. Under the influence of a chemoattractant secreted by the egg to complete fertilization, sperm chemotaxis towards the egg occurs, directing spermatozoa to it. These secretions may act not just by boosting motility or controlling the process, but also by activating motility that is not triggered by a change in ambient parameters following sperm discharge. In some situations, this secretion is contained in the seminal plasma and improves the motility and metabolism of spermatozoa. Ascidians and echinoderms, as well as fish like the Japanese bitterling or *Acheilognathus lanceolata*, herring (*Clupea pallasii*), rosy-barb (*Barbus conchonis*), and sea lamprey (Petromyzon marinus), have been observed to exhibit this phenomenon (Heinsbroek, 1991). When activated and exposed to these compounds, spermatozoa from these species behave differently in every situation. When sperm come into contact with good quality eggs in the right environment, their ability to form viable embryos is determined. Another issue with using fertilization experiments to determine semen quality is the lack of long-term outcomes. To be useful, quality assessment must be as accurate and quick as feasible. Fertilization rates could be determined several hours or minutes after fertilization, however in this case, just the cleavage activation is considered. The pace of embryos hatching is an indicator, yet development may take too long.

In aquaculture, egg quality has traditionally received far more attention than sperm quality. Generally, females have greater difficulty spawning than males, and issues in breeding are primarily attributable to egg failure, except in species where there are obvious challenges in spermiation or very poor sperm quality in captivity. Artificial fertilization using a big quantity of milt or pooled milt from several different species appears to have solved fertilization problems because poor quality samples are compensated with better ones. Nonetheless, this process is a waste of resources, and the use of an insufficient sperm/egg ratio may even decrease fertilization results. However, using many males does not guarantee that all of them will contribute to offspring development, because good milt producers will successfully fertilize the majority of the eggs, which could result in undesirable and unexpected inbreeding. As a result, an accurate assessment of sperm quality is of great interest in commercial aquaculture. Checking the quality of milt allows low-quality samples to be discarded, increasing the efficacy of artificial fertilization. The identification of those males with superior reproductive success is made possible by regular analysis of sperm generated by the broodstock. This is an important feature that, along with other genetic characteristics, must be considered when selecting appropriate broodstock males.

On the other side, sperm management in aquafarms frequently necessitates sperm storage for hours, days, or even years. The discovery of technologies for fresh storage at 4°C, as well as cryopreservation in liquid nitrogen at –196°C, has greatly aided sperm quality assessments. Cryopreservation is a complicated technique, and the quality of the sperm must be assessed after each step to determine the least harmful procedure. Only high-quality sperm samples may be properly cryopreserved, which has prompted sperm quality analysis. Many species of fish, mollusks, and crustaceans have been studied for sperm cryopreservation, and the field of sperm quality assessment, as well as procedures for sperm management and artificial fertilization, have benefited from this fact. Because sperm from aquatic species is extremely sensitive to contaminants, and the quality of milt from various species is frequently employed as a biomarker of water contamination, ecotoxicology has aided in the evaluation of sperm quality. These relationships, however, may vary depending on the species or experimental conditions used. For routine milt examination, parameters linked to motility, particularly the rate of progressing cells, are often used. Nonetheless, an appropriate assessment should always be based on a combination of attributes, including general sperm parameters, some motility characteristics, and a spermatozoa status evaluation. The combination of characteristics chosen is determined by the evaluation's target, as well as the available equipment and experience. For ordinary sperm handling on a farm, quality assessment can be quite straightforward, but it should be much more exact for experimental purposes. It's also important to remember that semen isn't a homogeneous mixture of cells and plasma, but rather a pool of cells derived from various spermatogonia, whose spermatogenesis may be asynchronous, resulting in a heterogeneous population of cells with different genotypes, maturation stages, and characteristics. This is why new sperm evaluation trends stress the examination of spermatozoa subpopulations for various determinations, raising concerns about the use of average sperm sample values.

A hemocytometer, a Neubauer chamber or any other counter chamber, and a phase contrast microscope are commonly used to count the cells in a known volume of milt. To get an adequate concentration of cells in the observation field, the sperm must be diluted at a predetermined pace before counting. Because of the viscosity of the semen, it is often difficult to get a uniform dilution of the sample, scores in the same sample can vary greatly depending on the aliquots pipetted for observation. On each sample, three or more readings must be taken, extreme values must be removed, and the average of numerous measures must be used as the result. This method is time-consuming and can be substituted with the Coulter counter, an electronic particle counter. The examination of the optical density of milt using a spectrophotometer is a faster and more reliable procedure. Optical density is unique to each species and fluctuates according to cell concentration. By centrifuging sperm to separate cells from plasma and then mixing them at varied rates, a standard curve of cell concentration can be created. After evaluating cell density with a hemocytometer, optical density in each sample with a varied concentration is assessed with a spectrophotometer. Cell density evaluation is quick after the standard curve is established. Filling special glass capillaries for hematocrit with the sample and centrifugation to separate the cells from plasma is how spermatocrite is measured. Spermatocrite is the ratio of cell volume to total sperm volume multiplied by 100. If specified equipment is not available, other tiny tubes can be utilized for testing. This measure is also a direct indicator of cell density, and it coincides with counting chambers.

Seminal plasma mineral components have been recognized as key factors influencing sperm quality. Osmolality, pH, and various spermatozoa activities, such as motility activation, are all affected by their presence or absence. In the seminal plasma of the majority of the species investigated, five ions predominate: sodium, potassium, chloride, calcium, and magnesium. Their concentrations will vary depending on the species, but for each ion, there is a reasonable range that provides the optimal environment for spermatozoa. Changes in this concentration could suggest a variety of spermatozoa disturbances. Contamination with blood, feces, or urine can alter the composition of these ions, and a low or high concentration of a certain ion can signal the start or end of the reproductive season, indicating worse sperm. However, for the same species, ion composition and concentration can vary significantly, making it challenging to interpret the data. Triglycerides, glycerol, fatty acids, glucose, malate, and lactate, for example, are all indicators of energy metabolism. Others, such as vitamins E and C, selenium, and citric acid, have a high anti-peroxidation impact and prevent oxidative processes and aging during storage in the reproductive tract. Vitamins, particularly vitamin C, cannot be synthesized by the animal and must be obtained from food, therefore their absence or low quantity in seminal plasma can suggest a nutritional problem with the broodstock, which is usually related with the use of frozen food. Low levels of vitamin C in rainbow trout seminal plasma were linked to lower sperm concentration, motility, and fertilizing capacity. Ascorbic acid's anti-oxidative properties have also been linked to spermatozoa plasma membrane integrity. Sugars, lipids, free amino acids, and proteins, among other seminal plasma elements, may also determine sperm quality. Plasma also contains lytic enzymes. Some of them, such as acid phosphatase (AcP), alkaline phosphatase, D-glucuronidase, or proteases, are responsible for removing degenerated spermatozoa at the conclusion of spawning, while others leak out of spermatozoa, signaling cell lysis or aging. The enzymes lactate dehydrogenase (LDH), aspartate aminotransferase (AspAT), and AcP are the ones most often linked to lytic processes and cell viability.

In the discipline of seminology, structural and ultrastructural studies provide more information. The structural pattern, shape, or size of spermatozoa can indicate a variety of characteristics, ranging from the species' evolutionary position to the cell's maturation stage or injuries induced by environmental causes. Complete descriptions of aberrant morphological patterns exist in animals, and structural traits such as the presence of cytoplasmic drops in the skull base are linked to defective epididymal maturation. Low fertility has also been linked to abnormal spermatozoa morphometry in many animals such as the bull, boar, and stallion. Several structural investigations in aquatic species are descriptive and focus on understanding taxonomic and evolutionary links. The investigation of sperm morphology and structural alterations of spermatozoa has also been focused at the evaluation of damage caused by cryopreservation or the action of pollutants. Pollutant-induced sperm abnormalities have been linked to decreased motility and fertilization capacity, as well as changes in plasma membrane, mitochondrial swelling, and tail morphology with changes in motility patterns and decreased fertilization ability after freezing and thawing. In comparison to mammals, however, data on the relationship between structural or morphometric traits and fertilization ability is scarce.

Different morphological patterns of sperm exist among aquacultured species, ranging from mammalian-like cells of fish, with a head containing the acrosome and nucleus, a well-developed mitochondrial sheath in the mid-piece, and one or two flagella, to the most primitively-shaped aquasperm of the majority of teleosts, in which structure is very simple. As a result, the information offered by structural analysis is of varying value. The more complicated spermatozoa are, the more cellular structures are required to maintain their order, and structural studies yield more information. Visual inspection using light microscopy is the simplest morphological evaluation, however the information available is limited due to the tiny size of most aquatic spermatozoa. This approach can be used to determine acrosome status, as well as acrosome reacting cells, morpho-anomalies, and spermatozoa subtypes. TEM and SEM are commonly used. Fine structure provides additional information, and changes such as chromatin de-condensation, mitochondrial enlargement, the production of blisters in the plasma membrane, changes in the shape of the tail or insertion into the head, and changes in the axonemal pattern can all be seen. Sample preparation must be done with extreme caution to minimize structural artifacts that make interpreting the data challenging, especially for TEM. Electron microscopy only provides incomplete information, and the methods are costly and time-consuming.

Because reproductive success is dependent on the quantity and/or quality of eggs generated during spawning, the quality of eggs can have a significant impact on fish aquaculture productivity and recruitment. In recent years, researchers have worked to develop accurate measures for identifying high-quality eggs. Although eggs have intrinsic qualities that influence egg quality, such as maternal age, genetics, yolk composition, size, and morphology, there are several extrinsic factors that influence egg quality throughout the spawning season, and thus larvae quality, such as ripening, manipulation, incubation conditions, and water quality, to name a few. Unfertilized or recently fertilized eggs are tested for quality markers, which are biological factors. They are used to assess the percentage of eggs that successfully complete one of the previously stated ontogenetic phases, as well as to discriminate between viable and nonviable eggs. Although fry production for key commercial freshwater fish has advanced to a high level of sophistication and is now a normal farming technique, egg quality can vary significantly, owing to variability caused by individual female conditions and overripening processes. Fry production for cultured marine fish is still more complex, yields are more unpredictable, and the process is limited to a few specialist hatcheries. Despite the ability of broodstocks to produce large quantities of eggs, egg quality frequently varies greatly and in an uncontrollable manner. Furthermore, viability and stress resistance in larvae coming from high quality egg batches may be better than in larvae originating from low quality egg batches. As a result, egg quality control in marine species is still more crucial than in freshwater species.

Earlier studies used empirical research to produce credible egg quality criteria for numerous species, including egg diameter, morphology, buoyancy, fertilization rate, and hatching rate; however, these characteristics were insufficient and/or inappropriate for evaluating spawn quality (Kamler, 2005). Researchers have established more precise metrics to identify egg quality in recent decades, and factors such as symmetry of early cell formation, yolk composition, critical intermediate metabolism enzymes, and molecular techniques have begun to be utilized to characterize egg quality. However, while these metrics can provide more conclusive information on egg quality, they are expensive and time-consuming. The quality of rearing fish is still exceedingly varied, with significant and changing levels of mortality, uneven growth, skeletal deformities, and aberrant coloration in flatfish. In reality, captive fish eggs are frequently inferior to wild fish eggs, implying that contemporary husbandry practices still have a significant impact on egg quality. Furthermore, little is known about how egg quality varies within or among females, or among populations during spawning season, which could be useful information for optimizing egg collection and larval production in commercial hatcheries; and whether a parameter is reliable for different species and types of eggs.

9.7. EGG QUALITY MANAGEMENT

Egg quality management technologies would allow broodstock maturation processes and conditions to be improved, and thereby fry production. From a commercial standpoint, these technologies would be a tool for ensuring product quality and keeping hatcheries running smoothly. Fish eggs, like those of other vertebrates, are generated as a result of an external interaction between eggs and sperm, albeit this group has unique reproductive processes. Despite the fact that both gametes contribute to egg quality, studies were primarily concentrated on the maternal contribution, because females are responsible for the synthesis of the yolk and other components required for optimal larval development. Nonetheless, various research has looked into the role of males and sperm quality in egg quality in recent years. Both paternal and maternal variables contribute to total egg/larva viability, although at different periods and in different ways. In reality, paternal effects appear early in ontogenesis, affecting spermatozoan density and motility, whereas maternal influences appear later, affecting embryonic survival and responding to egg ripeness and female age through egg matter composition. Pelagic and demersal eggs are two types of fish eggs. The stage of larval development at hatching is the most important difference between these two types of eggs, because pelagic eggs hatch into buoyant blind larvae with the mouth and anus closed, whereas demersal eggs hatch into eyed, freely swimming larvae with the ability to feed on formulated diets.

Because demersal eggs have a longer incubation period than pelagic eggs, several development phases are identified during embryonic development, and external influences will affect larval quality considerably and differently in these types of fish eggs. As a result, quality metrics for demersal and pelagic eggs may differ. This heterogeneity has no bearing on hatching, but it does make determining egg quality more difficult because the investigated characteristics may fluctuate between individuals and developmental phases. As a result, quality indicators must remain consistent throughout the development stages essential to determining egg quality, or variations during development must be at least as small as those caused by quality disparities. To summarize, egg quality assays must be designed for each species and for each manner of egg collecting. Individual egg samples are taken for species where gametes can be removed. The ovarian fluid and eggs are separated, and the eggs are cleaned to remove any ovarian fluid residues. Different spawns are collected from the spawning tanks through egg collectors for species whose gametes are acquired by natural spawning. The samples are then separated into subsamples for extraction and fixation operations, followed by analysis. To determine embryo or larvae viability criteria, one sub-sample of each egg sample is hatched. Ovarian fluid is examined directly. Finally, correlation coefficients, ANOVA models, and regression models are used to assess statistical correlations between analyte levels and viability parameters.

Furthermore, larger eggs typically have a higher yolk content, which allows hatched larvae to live in the wild for longer periods of time without feeding, boosting their chances of survival. However, experiments with rainbow trout, Oncorhynchus, Atlantic salmon, turbot, and Senegalese, all of which were kept in favorable conditions, revealed that egg size was not a decisive factor for larval survival and growth rate. When it comes to pelagic eggs, where a single diameter is often employed, egg size is usually computed based on egg diameter. However, two diameters are occasionally mentioned: the longest, egg length, and the second, egg breadth, which is perpendicular to it. Individual batch-spawning females may experience a decrease in egg diameter and hence yolk content near the conclusion of the spawning period, which has been linked to the exhaustion of vitellogenesis reserves and may result in a fall in the embryo and larvae survival percentage. Fish age appears to be less relevant in influencing fertility than fish size, with observed increases in egg number with age most likely related to concurrent increases in fish size.

Following fertilization, the cytoplasm is partitioned into many cells called blastomeres by a rapid series of mitotic cell divisions. The discoidal meroblastic cleavage pattern is found in fish eggs, and it restricts cell cleavage to the animal pole. At the 8-cell stage, eggs are individually analyzed for five characteristics: symmetry, uniformity, margins, inclusion bodies, and cell adhesion; the operator evaluates each egg on a qualitative scale of 1 (extremely abnormal) to 4 (normal). Using blastomere morphology as a quality criterion offers a significant cost benefit because it is possible to detect and reject poor quality spawn a few hours or days after fertilization, saving energy, space, and man hours. Furthermore, the majority of abnormalities arise after fertilization, resulting in egg abortion if egg repair mechanisms fail to correct the identified deformities. Asymmetric blastomere organization, disparity of blastomere size, poor adhesion between blastomeres, and poor delineation of blastomere borders were all classified as abnormalities in blastomere cleavage. Fish have a wide variety of reproduction methods. The fusion of male and female gametes generated from individuals of opposite sexes in the process of external fertilization is the most common pattern of reproduction found in most farmed fish species. However, there are various reproduction techniques that deviate greatly from the pattern depicted.

Fish use a variety of fertilization methods, including external fertilization. Even oviparous fish are divided into two groups: ovuliparous and zygoparous fish. Ovuliparous fish release mature oocytes for external fertilization, whereas zygoparous fish lay embryos after internal fertilization. Internally fertilized fish are divided into two groups: ovoviparous fish, which incubate their embryos in a modified part of the female's oviduct, and viviparous fish, which develop fertilized eggs in the ovaries or the uterus. Finally, certain species practice asexual reproduction, in which male sperm is used simply to trigger the development of eggs but does not contribute genetically to the developing embryo. Only female individuals occur in numerous species that fertilize with sperm from related species. In the Amazon molly, viviparity and all-female asexual reproduction are observed. Hydrophilic polysaccharides, which bind to water and hence speed up water influx, are found in cortical alveoli. This inflow causes the egg envelope to enlarge, which varies depending on the species. The swelling of eggs in common carp eggs is about 3–6 times that of grass carp eggs, which can be as much as 50–100 times. Cortical alveolar excretions react with the egg membrane as well. The egg membrane hardens as a result of this process and serves as a major barrier for the growing embryo from chemical and physical interactions. It allows oxygen-rich water to enter and toxic metabolic wastes to be excreted. The thickness varies according to the species. Pelagic eggs often have a thinner fertilization envelope than sticky eggs. During activation, the egg's sticky coating also forms. It can extend across the entire egg surface or only a small portion of it. The adhesive layer forms a ring around the animal pole in the African catfish, and the micropyle is placed in the center of this ring.

Egg activation happens without the connection of sperm and egg, hence eggs immersed in water will eventually become activated. However, it has been discovered in numerous species that these responses occur much later in unfertilized eggs than in fertilized eggs. Chemotaxis attracts spermatozoa to fish eggs, albeit its exact purpose is yet unknown. Several investigations have indirectly demonstrated the presence of a chemoattractant. The spermatozoa of the rosy barb or *Barbus conchonius* are not only guided by the special grooves and ridges on the egg surface around the micropyle, but the majority of sperm cells arriving in the first 30 minutes post-activation preferentially choose this guidance system. After water hardening, however, the efficacy of this mechanism is diminished, and only a small percentage of spermatozoa use the guidance system 80 minutes after fertilization. In most species, the micropylar canal is large enough for only one spermatozoon to travel through at a time. The common carp micropyle is an anomaly, as its internal aperture is large enough to admit two cells simultaneously. A fertilization cone develops and extends into the micropyle as the head of the fertilizing spermatozoon fuses with the sperm entrance site of the egg membrane. The fertilization cone reaches a large size in certain species, and its tip is visible in the micropyle's external aperture. The fertilizing spermatozoon's flagellum is still visible outside the fertilization cone at this point.

The occurrence of polyspermy occurs when more than one spermatozoon fuses with the sperm entrance point. Polyspermy is undesirable because polyspermic embryos do not survive development and perish. Polyspermy in fish is prevented by a variety of methods. The micropyle of most species' eggs is wide enough to accommodate only one spermatozoon. Spermatozoa can only fuse with the egg membrane at the sperm entry site, and their diameter is similarly limited to one sperm cell's size. The immediate development of the fertilization cone and its protrusion into the micropyle prevents subsequent spermatozoa from entering the pycropylar canal and forces supernumerary sperm cells out. Finally, supernumerary spermatozoa in the perivitelline area might be agglutinated by the colloids secreted from the cortical alveoli. In sturgeons, the specific mechanism of acrosome response is still unknown. In other species, the acrosome includes enzymes that help the fertilizing spermatozoon pass through the egg's zona pellucida. Fish eggs, on the other hand, have a micropyle that connects directly to the egg membrane, eliminating the need for an acrosome reaction. Theoretically, this arrangement is negative since it encourages polyspermy. In sturgeon, polyspermic fertilization can be generated, but it is uncommon in natural reproduction. Polyspermy is rare due to a number of variables. Cytoplasmic processes project into the micropyles from all sperm entrance sites after penetration of the fertilizing spermatozoon. In these species, rapid exocytosis of cortical alveoli is also critical in preventing polyspermy.

INDEX

A

Acheilognathus lanceolata 202
acid phosphatase (AcP) 205
advertising 60, 66, 68
aerosols 152
agriculture 152, 157, 159
Algal production 79
alkaline phosphatase 205
alkalinity 178, 182
ammonia 178, 179, 183
Animal manure 30
aquaculture business 56, 58, 59, 61, 62, 68
aquaculture community 93, 105
aquaculture endeavors 82
aquaculture foods 28
Aristichthys nobilis 91
artificial feed 10
aspartate aminotransferase (AspAT) 205
asset-generating loans 69
Asterina pectinifera 201
Atlantic salmon farming 29

B

Bacterial diversity 123
baking soda 178
balance sheet 70, 71
beef 122
benthic enrichment 123
best management practices (BMPs) 123
Best management practices or BMPs 94
bighead carp 32
biodiversity 90, 98
biofilter 79, 81
biomass 9, 19
Brachionus angularis 87
Brachionus falcatus 87

C

cage system 10
carbon 30, 34, 38, 42, 45, 47, 48, 50, 52
carbon dioxide 178, 179, 180, 181, 184, 190
Cash flow 72
Cash revenue 5
cellular pathways 155

Chirostoma estor 127
Chlorine 180
Cichlasoma urophthalmus 127, 128
Climate change 152, 156, 162, 167, 171, 172
Climate-related parameters 155
Coastal crops 112
coastal erosion 157, 158
coastal flooding 157
coastal systems 157
combined intensive-extensive (CIE) system 79
common carp 7, 14, 15, 17, 32, 42
Copepoda nauplii 87
copepods 78, 87
crayfish 105, 109, 110
Credit 71
Credit reserves 71
Crustaceans 78
Ctenopharyngodon idella 91
cultivation 2, 10, 17, 19, 20, 21
cultural organisms 10
Cyprinus carpio 91, 125

D

detritivorous 83
D-glucuronidase 205
dilute metabolic wastes 32
disease 123, 125, 139, 143, 145, 146, 147, 148
disease control 28
Dissolved oxygen (DO) 86

E

Ecology 155
ecosystem services 98
egg 192, 193, 194, 195, 199, 200, 201, 202, 203, 206, 207, 208, 209, 210, 211
El Nino Southern Oscillation (ENSO) 113
embryonic development 192, 208
energy 57
Environmental implications 93
environmental management 95
erosion 123
Eutrophication 123
Evapotranspiration 110, 111
extreme weather events 157

F

Filinia terminalis 87
fish culture 2, 16
fish farming 2, 3, 4, 5, 6, 8, 13, 15, 16, 21, 22, 24
fluorine 180
food chain 83
food consumption 152
food security 4, 122, 149
freshwater aquaculture land 7
Freshwater aquaculture resources 6
fully recirculating systems 178

G

geology 180
germ cell 194
Girardinichthys multiradiatus 129, 131
global climate change 90
global ecosystems 93
Goodea atripinis 129, 130
Goodea gracilis 129, 131
granulosa cells 194
grass carp 7, 32
greenhouse gas 99
Groundwater 180, 190

H

Habitat degradation 154
habitat modification 123
Haliotis diversicolor 197
Herbivorous 83
Herinchthys cyanogutattus 127, 128
herring 201, 202
High-risk loans 69
hydrogen sulfide 180
Hypophthalmichthys molotrix 91

I

Ictalurus 105, 119, 200
Ictalurus punctatus 105
Indoor aquaculture 176
inorganic fertilizers 30
intensive aquaculture 28, 36, 37, 40, 47
intensive tilapia farming 76
Intergovernmental Panel on Climate Change (IPCC) 113
iodine 122

L

Labeo calbasu 7
lactate dehydrogenase (LDH) 205
liquid waste 123
Litopenaeus vannamei 111
livestock production 30
loans 69, 70, 71

M

Macrobrachium rosenbergii 129, 132
Macrozoarces americanus 197, 198
magnesium 122
mariculture 152, 157
marine aquaculture 152, 154
Marine capture fisheries 152, 153
marine fisheries 90
marine resources 90
marine shrimp 61
marketing plan 59, 64, 66, 67, 68
market research 66
market systems 4
methane 180
Modern aquaculture 28
molluscophagous 83
Mugil cephalus 125, 126

N

natural aquatic systems 91
natural food 30, 31, 32, 34, 43, 51
natural habitat 98
Natural oyster reefs 122, 123
natural sources 2
nematodes 78
niche market 62, 67
Nitrification 79, 80, 81
nutritional content 123, 138

O

ocean acidification 157, 158, 161, 171, 172
Ocean warming 157
ODAS (organic detrital algae soup) 184
omega-3 fatty acid 122
Oocytes 193, 194
Oreochromis niloticus 32
organic matter (OM) 83
ovary 193, 194
overfishing 154, 158, 173
ovigerous lamellae 193
oyster shellfish aquaculture 122

P

packaging 63, 67, 68
partial-reuse systems 178
Penaeus japonicus 35
Penaeus subtilis 42
Penaeus vannamei 76, 77
Phormidium tenue 87
phosphorus 122, 125, 145
physiology 155
Phytoplankton 30, 31
phytoplanktonophagous 83
polychaetes 78
Polyculture 83
Pond culture 176
pond ecology 78
pond food web 30
Post-ovulatory follicles or POF 194
Procambarus 105
Procambarus chrkii 83
Procambarus digueti 127
production costs 122, 134
proteases 205

R

rate of return 68
recirculating aquaculture systems (RAS) 94, 176
red swamp crawfish 83
reproductive cycle 192, 194
Reservoir development 9, 10
reuse systems 177, 178
revenue 58, 62, 69, 71, 72, 73
rosy-barb 202

S

saltwater intrusion 157
Salvelinus niger 119
Seafood 122
sea lamprey 202
seedstock 61
selenium 122
Self-liquidating loans 69, 71
semi-intensive pond farming systems 30
Serial-reuse systems 178
Sertoli cells 195
Silver carp 32
Single cell proteins (SCP) 37
Sodium ions 201
Spawning 192, 201
Spermatozoa 201, 211
static pond culture 9
static water 9
stratosphere 152
Submergence 157
Supplementary feeds 31
Supplementary foods 31
supply chain 59, 67, 68

T

thelycum 199
thelycum penaeoideans 199
Tilapia zilli 32
Total ammonium nitrogen (TAN) 86
Total suspended solid (TSS) 86
toxicity 178, 179, 181, 182
transportation 63, 68

U

ultraviolet (UV) radiation 189
United States Environmental Protection Agency (USEPA) 101

V

viruses 123, 148
Vitamins 76, 78

vitamin supplementation 76, 78

W

Water flow 183
water quality 123, 124, 129, 134, 143, 146, 148
water quality control 176
water quality management 28
water-soluble vitamins 76
weather pattern changes 157

X

Xenotoca variata 129, 130

Y

yeasts 37

Z

zooplanktonophagous 83